# Fretting Friction and Wear at Bolted Joint Interfaces

**Online at: https://doi.org/10.1088/978-0-7503-6214-6**

# Fretting Friction and Wear at Bolted Joint Interfaces

**Chao Xu**
*School of Astronautics, Northwestern Polytechnical University, Xi'an, Shaanxi, China*

**Dongwu Li**
*School of Astronautics, Northwestern Polytechnical University, Xi'an, Shaanxi, China*

**IOP** Publishing, Bristol, UK

ISBN 978-0-7503-6214-6 (ebook)
ISBN 978-0-7503-6212-2 (print)
ISBN 978-0-7503-6215-3 (myPrint)
ISBN 978-0-7503-6213-9 (mobi)

DOI 10.1088/978-0-7503-6214-6

Version: 20241201

IOP ebooks

British Library Cataloguing-in-Publication Data: A catalogue record for this book is available from the British Library.

Published by IOP Publishing, wholly owned by The Institute of Physics, London

IOP Publishing, No.2 The Distillery, Glassfields, Avon Street, Bristol, BS2 0GR, UK

US Office: IOP Publishing, Inc., 190 North Independence Mall West, Suite 601, Philadelphia, PA 19106, USA

# Contents

# Preface

Bolted joints are a widely used mechanical fastening method, found in various machines, automobiles, aerospace systems, and architectural structures. Bolted connections offer strong preload along with the advantages of easy disassembly and low cost. However, a small amount of relative motion at connection interfaces, known as 'fretting', can lead to wear, preload loss, and nonlinear mechanical behavior, which can significantly impact the service life of equipment and increase the difficulty in the design and control of the connection structures.

Although fretting friction and wear are common phenomena in many mechanical joints, the underlying mechanics and interactions are complex and present significant challenges. Fretting-induced wear not only impacts the structural integrity of bolted joints but can also lead to system failure, especially in vibration or impact environments. As industrial technologies continue to evolve, understanding and controlling the effects of fretting at mechanical joint interfaces has become a critical concern for both engineers and researchers.

This book aims to provide readers with the latest research on fretting friction and wear problems at bolted joint interfaces. It not only presents the experimental findings related to fretting friction and wear but also explores the latest advancements in numerical simulation techniques and their applications. These insights are valuable not only for researchers in mechanical design and structural health monitoring but also offer practical guidance to engineers working in relevant industrial fields.

It is our hope that this book will serve as a valuable resource for researchers and engineers exploring the mechanics of joint structures, inspiring further innovation and practical applications in this critical area.

The authors

# Acknowledgments

The authors would like to acknowledge support from the National Natural Science Foundation of China (Grant No. 12472043, 12202264, 12072268).

# Author biographies

## Chao Xu

**Chao Xu** received his PhD degree in aerospace Science and technology from Northwestern Polytechnical University, Xi'an, China, in 2007 and is currently a professor in the School of Astronautics, Northwestern Polytechnical University. His research interests include structural health monitoring, structural dynamics of aerocraft and aerocraft structure design.

## Dongwu Li

**Dongwu Li** is currently an associate professor at Northwestern Polytechnical University, Xi'an, China. He received BS, MS, and PhD degrees in aerospace engineering from Northwestern Polytechnical University in 2014, 2017, and 2021, respectively. His main research interests include structural dynamics and contact mechanics.

# Symbols

| | |
|---|---|
| $t$ | time |
| $\Delta t$ | time interval |
| $\delta$ | tangential relative displacement |
| $R_a$ | surface roughness |
| $f$ | excitation frequency |
| $N$ | bolt preload |
| $N_0$ | initial value of bolt preload |
| $T$ | friction force |
| $k_t$ | tangential contact stiffness |
| $\mu$ | friction coefficient |
| $\Delta x$ | nominal displacement of piezoelectric actuator |
| $U$ | input voltage of piezoelectric actuator |
| $C$ | electrostatic capacity of piezoelectric actuator |
| $k_b$ | bending stiffness between the bolt head and the nut |
| $L_s$ | bolt shank length |
| $E_s$ | elastic modulus of the bolt shank |
| $I_z$ | the second moment of area of the bolt shank |
| $G$ | shear modulus |
| $\Delta F$ | correction parameter of friction force |
| $\Delta A$ | average sliding stroke |
| $P_r$ | relative reduction of bolt preload |
| $\vartheta$ | Poisson's ratio |
| $a$ | contact radius |
| $f_0$ | central frequency of random process |
| $\sigma$ | standard deviation |
| $n$ | number of Jenkins element |
| $f^*$ | critical sliding force |
| $\varphi$ | probability density function |
| $\Delta$ | range of critical sliding force |
| $\delta_m$ | amplitude of tangential relative displacement |
| $T_m$ | amplitude of friction force |
| $p$ | contact pressure |
| $p_0$ | contact pressure at contact center |
| $p_{\mathrm{eq}}$ | equivalent contact pressure |
| $\tau$ | tangential sliding stress |
| $\tau_0$ | contact pressure at contact center |
| $E$ | Young's modulus |
| $\gamma$ | proportional coefficient |
| ub | upper bound of integral |
| lb | lower bound of integral |
| $m$ | mass |
| $k$ | linear stiffness |
| $j$ | order of Fourier series |
| $n_h$ | truncated order of harmonics |
| $\boldsymbol{f}_{\mathrm{nl}}$ | nonlinear force vector |
| $\boldsymbol{M}$ | mass matrix |
| $\boldsymbol{C}$ | damping matrix |

| | |
|---|---|
| $\boldsymbol{K}$ | stiffness matrix |
| $\boldsymbol{F}_e$ | external excitation vector |
| $\boldsymbol{u}$ | displacement vector |
| $r$ | distance to bolt hole center |
| $z$ | rough surface height |
| $D$ | fractal dimension |
| $H$ | hardness of material |
| $\sigma_y$ | yield strength of material |
| $P_{\text{sd}}$ | power spectral density |
| $E_d$ | cumulative dissipated energy |
| $N_c$ | wear cycle number |

# Chapter 1

## Introduction

This chapter introduces the background and significance of bolted connections, and reviews the current research progress in this field from three perspectives: interface fretting test technology, friction and wear behavior at joint interfaces, and contact modeling of joint interfaces.

### 1.1 Background

Bolted connections are commonly used in various mechanical systems, such as turbine engines, high-speed trains, spacecraft, and wind turbines, as one of the most prevalent and crucial fasteners for transmitting loads and maintaining the structural integrity of the system [1]. For example, the Chinese CZ-7 launch vehicle has around 680 000 fasteners and the next generation of Chinese manned rockets is expected to have millions of fasteners, which account for approximately 10%–16% of the total weight of the spacecraft cabin.

Failure of bolted connections can lead to severe equipment damage and, in some cases, even catastrophic accidents, resulting in personal injury and economic losses [2]. Many reported accidents are caused by bolted connection failures. For example, in 1979, American Airlines Flight 191 lost an engine in flight due to issues with the engine hanger fixing bolts [3]; in 2014, Porsche 911 GT3 vehicles caught fire due to loose engine bolts, prompting a recall of 785 cars to replace the engines [4]; Tesla Inc. discovered that bolts in the steering system of 2016 Model X vehicles were prone to corrosion in highly corrosive environments, leading to a recall of over 3000 vehicles [5]; and, in 2024, an American Airlines Boeing 737 MAX 9 had an emergency door detach shortly after takeoff due to loose fastening bolts [6]. During the operation of mechanical systems, bolts not only need to withstand enormous loads, but also must maintain reliability over the long term under complex dynamic conditions. Therefore, the strength, stiffness and life design of bolted connections are crucial to overall structures [7]. Croccolo *et al* [8] summarizes the failures involving threaded fasteners in the literature and categorizes the failures

doi:10.1088/978-0-7503-6214-6ch1

based on their root causes, such as incorrect assembly, overloading, fatigue, and loss of preload during operation.

Although bolted connections have the advantages of high reliability and ease of assembly and disassembly, the presence of contact interfaces introduces discontinuities and nonlinearities in assembled structures. These inherent nonlinearities of joint interfaces caused by contact significantly affect the dynamic response of the assembly, which results in nonlinear dynamic behaviors, such as amplitude-dependent properties, sub/super harmonics, and instabilities, complicating the prediction of dynamic response and the control of assembled structures. On the other hand, friction hysteresis at connected interfaces can dissipate vibration energy, this mechanism thus is also used to reduce vibration in some mechanical systems, such as dry friction dampers in aircraft turbine engines [9–12].

The friction contact at bolted joint interfaces is highly dependent on external excitation. As the excitation amplitude varies, the relative motion between the interfaces spans multiple scales ranging from partial slip, on the order of micrometers or smaller, to gross slip across the entire joint surface. This requires taking sufficiently small steps in the time domain to accurately capture the interface motion, resulting in high computational costs for calculating the dynamic response of real assembled structures. In addition, the frictional contact at connected interfaces exhibits certain evolutionary characteristics over time, which involves changes in surface morphology, redistribution of contact pressure, decay of bolt preload, changes in contact parameters, etc [13, 14].

A better understanding and accurate quantification of the mechanics behavior of joint interfaces would contribute to analyzing the dynamic characteristics of bolted joint structures, enabling engineers to design mechanical structures with high performance and efficiency. Therefore, it is extremely important to improve the understanding and characterization of fretting friction and wear behavior at joint interfaces, which has become a pressing need in several industries such as aerospace, automotive, and rail transportation.

## 1.2 Interface fretting test technology

Experimental studies on friction and wear at connection interfaces help to understand the mechanical behavior of contact surfaces, verify friction contact models, and estimate the contact parameters needed for the application of these models. Early experiments on the nonlinear behavior of connection structures can be traced back to the mid-nineteenth century, focusing mainly on the nonlinear damping behavior of connected structures through the study of structural vibration responses [15–18]. Ungar *et al* [16] experimentally analyzed the dynamic response and dissipation mechanisms of connection damping in riveted skin structures and lap joint beams in spacecraft, and proposed a method for estimating connection damping. Rogers *et al* [17] conducted experimental studies on the energy dissipation characteristics of planar contact under tangential oscillating loads, analyzing the effects of load amplitude, interface conditions, and materials on connection damping, and suggested an approach for vibration control by adjusting the load. These experiments primarily explored the energy dissipation mechanisms at connection interfaces, revealed that connection damping differs from proportional damping, and discovered the amplitude-dependent energy dissipation. These findings lay a foundation for the study of interface friction and dynamics of connected structures.

With the rapid development of experimental techniques, an increasing number of studies are adopting direct measurement methods to investigate the mechanical behavior between joint surfaces. Based on the way the external load is applied, these fretting experimental methods can be categorized into quasi-static and dynamic methods. Quasi-static methods often use servo-hydraulic fatigue testing machines to provide tangential excitation. Dynamic methods are based on vibration testing principles, usually using electromagnetic shakers. The wear test duration in dynamic experiments is significantly shorter than in quasi-static tests, but the robust control of the system is more challenging.

Gaul *et al* [18, 19] proposed a bolted joint fretting test device based on the principle of structural resonance for interface fretting friction of bolted connection, as shown in figure 1.1(a). The friction hysteresis characteristics and nonlinear dynamic response of typical bolted joint interfaces were investigated. This device uses an electromagnetic shaker to induce the desired longitudinal vibration mode, and then measures the acceleration response of the bolted joint components through accelerometers. The inertia force of the free-end mass block is considered the friction force transmitted through the joint interface. In [18], the device was also used to study the shift of structural resonance peaks under different load conditions, hysteresis curves, and the effects of surface fretting wear on the structural frequency responses.

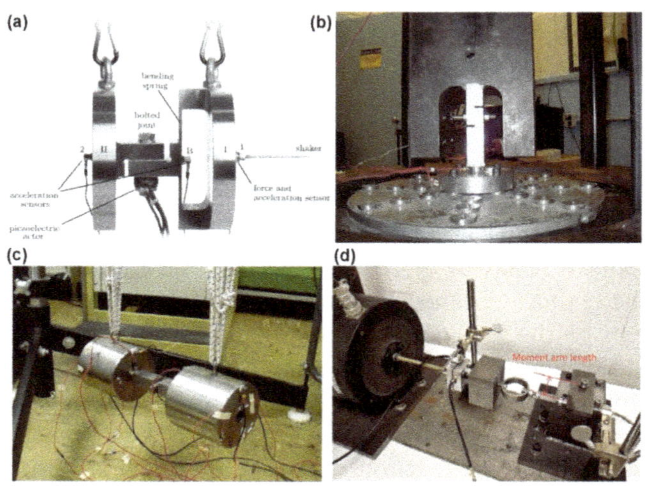


**Figure 1.1.** Typical dynamic testing systems for measuring fretting friction at bolted joint interfaces: (a) Gaul resonator, (b) Sandia single degree-of-freedom device, (c) Sandia dumbbell experimental set-up, and (d) Test rig for determining friction damping in bolted structures. ((a) Reproduced with permission from [19]. Copyright 2015 Elsevier. (b) and (c) Reproduced from [20]. Public domain. (d) Reproduced with permission from [21]. Copyright 2018 Springer Nature.)

Sandia National Laboratories [20] developed a simple, fixed, single degree-of-freedom experimental device, including a large rigid mass, as shown in figure 1.1(b). The system was loaded with a controlled sinusoidal excitation and the acceleration response was measured. The measurements are conducted at the resonance of the device. They found that the energy dissipation due to microslip (also called partial slip) is a nonlinear mechanism and the joint stiffness is also nonlinear. Furthermore, the results show that the energy dissipated per cycle of the contact surface has a power-law relationship with the load amplitude, and the power exponent is less than 3. Later, this device was extended to a simpler and cleaner set-up with a dumbbell configuration, as shown in figure 1.1(c). The bolted connection specimen to be tested was placed between two heavy dumbbell blocks suspended by flexible ropes. The external dynamic excitation is applied on the end of one of the blocks and triggers the first axial mode of the system. The overall layout of this device is very similar to the Gaul resonator.

In the above three test apparatuses, the displacement response of the joint interface was obtained using time domain or frequency domain integration methods to integrate the directly measured acceleration signal. Strictly speaking, the dynamic test method can be regarded as an indirect technique in which the energy dissipation of joints is inferred according to the system response.

Sanati *et al* [21] proposed a new experimental approach to determine the friction-induced damping of bolted joints, in which the joint was isolated by adding a mechanical resonator consisting of a lumped mass and spring to the bolted joint structure, as shown in figure 1.1(d). The tangential force was measured using a force transducer and the displacements of the resonator and the bolted beam were measured by a fiber optic displacement sensor and a wide frequency capacitive sensor. However, the measured hysteresis curves are weird. The friction force is too small relative to the applied bolt preload (the friction coefficient inferred from this is very low and does not conform to reality), and the shape of the hysteresis loop is like an ellipse, which may be caused by the phase difference between force and displacement.

The advantage of the dynamic testing method is that it can simultaneously measure the friction hysteresis curve of the joint interface and the dynamic behavior of the connected structure, and can well reveal the nonlinear dynamics of the connected structure. However, unavoidable noise in the measured data may introduce errors during the post-processing integration, and the triggered interface sliding amplitude is limited (in some experiments no gross slip of interfaces can be observed). The errors introduced by above indirect measurements have been well avoided by some later devices. Although the tested samples of these devices are not all bolted joint specimens, the key technologies involved have contributed to the advancement of fretting friction and wear testing techniques for bolted joint interfaces. Typical representatives include the test rig developed by Padmanabhan *et al* [22], the first-generation [23] and second-generation [24] fretting test set-ups of Imperial College London, and the first-generation [25] and second-generation [26] fretting test rigs of Politecnico di Torino. These devices can directly measure the relative displacement and friction force at contact interfaces generally using laser vibrometers (or displacement probe) and force sensors.

In quasi-static fretting testing, external cyclic excitation is usually applied using a fatigue testing machine or a piezoelectric actuator. Abad *et al* [27] used a universal testing machine to perform quasi-static friction tests on bolted joints and obtained hysteresis curves under different loads, as shown in figure 1.2(a). In this device the bolt preload was directly measured using a force washer, which can achieve higher accuracy than the torque estimation method.

Eriten *et al* [28] developed a fretting apparatus to measure the hysteresis loops and contact parameters of bolted joints, as shown in figure 1.2(b), where a piezoelectric actuator is employed to provide oscillatory motions and a three-axis force sensor to measure the tangential friction force and to monitor the out-of-phase motion. They studied the influence of bolt preload, tangential displacement amplitude, and material on the joint parameters, and the effects of surface roughness and lubrication on hysteresis loops at the early stage of the fretting of bolted joints [29]. The device uses a closed-loop controlled module, has the characteristics of high reliability, and can provide broadband excitation. However, it ignores the deformation of the specimen at the fixed support end, and the displacement measurement point is far away from the contact area, resulting in measurement errors of the interface displacement.

Li *et al* [30] presented a novel fretting test rig to improve the measurement accuracy of hysteresis curves in bolted connection and study the friction and wear behavior at joint interfaces, as shown in figure 1.2(c). This apparatus allows the interface relative displacement to be measured using one laser beam and a prism.

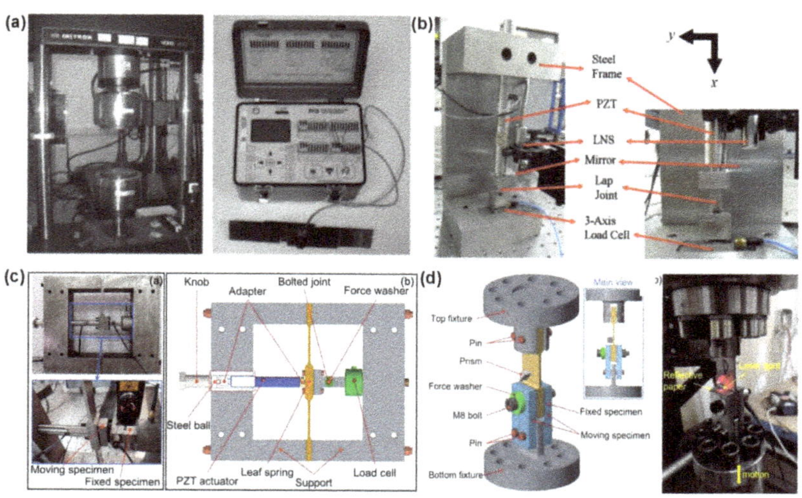


**Figure 1.2.** Typical quasi-static test rigs for measuring hysteresis curves at bolted joint interfaces developed by (a) Abad *et al* [27], (b) Eriten *et al* [28], (c) Li *et al* [31], and (d) Li *et al* [33]. ((a) Reproduced with permission from [27]. Copyright 2012 Elsevier. (b) Reproduced with permission from [28]. Copyright 2011 Springer Nature. (c) Reproduced with permission from [31]. Copyright 2020 Elsevier. (d) Reproduced with permission from [33]. Copyright 2022 Elsevier.)

A leaf spring is exploited to guide the motion of joint interfaces and avoid the out-of-phase motion, and a force washer is used to measure the bolt preload directly. The authors explored the wear behavior of bolted joint interfaces and the evolution of contact parameters under long-term sinusoidal and random excitation by this test rig [31, 32]. However, the load outputted by the piezoelectric actuator is limited, which brings the challenge of getting the hysteresis curves in a larger bolt preload. Furthermore, since the piezoelectric actuator was not equipped with a cooling system, the high temperature caused by long-term operation (in wear tests) potentially threatens the safety of the actuator.

Li *et al* [33] measured the fretting response of bolted joint interfaces under transversal vibration using a servo-hydraulic fatigue testing system in which the joint specimens are arranged symmetrically to avoid the influence of other contact surfaces besides the target surface on the measurement results, as shown in figure 1.2 (d). The fixed specimen is designed to have high flexibility, which is available for the formation of plane contact instead of a local contact (caused by additional bending moment). The results indirectly verify the origin of the residual stiffness frequently observed in experiments.

Additionally, some experiments are conducted on bolted joint interfaces under torsional cyclic loading [34, 35] where apparent hysteresis nonlinearity can also be observed. These works have been introduced in detail in a recently published review paper [36] and will not be elaborated here. In interface fretting tests, the measurement of bolt preload is also a key point. Bickford [37] summarized several basic methods for controlling the preload during bolted joint assembly, including torque control, torque and rotation control, and tension control methods. Although these traditional techniques allow indirect control of the preload during assembly, they cannot provide accurate values during operation. Typical measurement techniques for bolted joint preload include strain gauges, force washers, pressure membranes, and ultrasonic methods [38].

Although experimental techniques for bolted connections have made significant progress, the excitation forms and directions in current experiments are relatively singular, which differs from the complex conditions in actual engineering structures. Therefore, an important development direction in experimental techniques is to study the test technology of bolted connection under composite excitation, a topic that has gained considerable attention in recent years in industries such as aerospace engines, high-speed rail, and wind power.

## 1.3 Friction and wear behavior at joint interfaces

The connected components come into contact with each other through a certain normal preload method, relying on tangential friction at the interface to transmit load and energy. Under vibrational loads, intermittent separation and contact may occur in the normal direction perpendicular to the mating surface, while relative motion may occur in the tangential direction of the interface. In the 1950s, Mindlin [39, 40] studied the problem of tangential sliding in spherical contacts, suggesting that sliding in the contact area starts from the periphery and gradually expands inward as tangential excitation increases (related to the distribution of contact pressure). Based on Hertzian contact theory and the Coulomb law of friction, Mindlin derived an analytical expression for the relationship between tangential displacement and tangential force for elastic spheres under the combined action of constant normal load and oscillating tangential load. The author further derived that, at low amplitudes, the energy dissipation per cycle is proportional to the cube of the tangential load amplitude. Johnson [41] tested the friction behavior of sphere-to-plane contact under oscillating tangential load using a fretting test apparatus, and the experimental results confirmed Mindlin contact theory. The Lubrication and Wear Group *et al* [42] used photoelastic frozen stress techniques to analyze the tangential–normal contact relationship, confirming the proportional relationship between tangential force distribution and normal pressure.

With the deepening of theoretical and experimental research, understanding of the contact behavior at connection interfaces has become more comprehensive. Numerous friction experiment results indicate that frictional contact forces exhibit nonlinear characteristics [20, 26, 28, 43]. Under periodic excitation, the friction force and relative displacement at the interface form a hysteresis loop. Figure 1.3 shows a typical hysteresis curve and its physical meaning [44], which includes three states of surface motion: stick, microslip (also known as partial slip), and gross slip. (i) In the stick regime, the relationship between friction force and relative displacement is linear, and only elastic deformation occurs between the contact surfaces. (ii) In the microslip regime, relative sliding occurs in local contact regions while other regions remain in the stick regime, resulting in a clearly nonlinear relationship between friction force and relative displacement. (iii) In the gross slip regime, the sliding region expands to the entire contact surface, the friction force reaches its maximum value, and it almost no longer changes with relative displacement (in some cases, it changes linearly, indicating the presence of residual stiffness). The area enclosed by

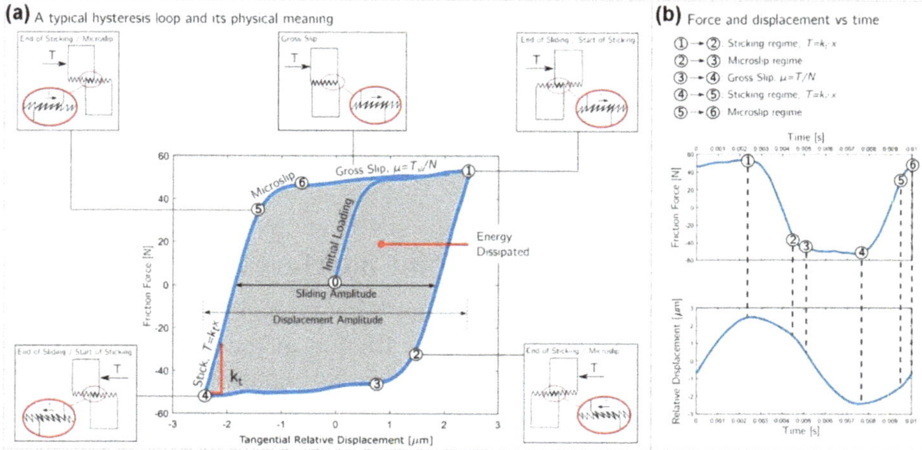


**Figure 1.3.** (a) A typical hysteresis loop and its physical meaning and (b) friction force and relative displacement versus time. (Reproduced from [44]. CC BY 4.0.)

the hysteresis curve represents the energy dissipated by the interface friction mechanism.

The hysteresis friction behavior between contact surfaces has a significant impact on the dynamic characteristics of connected structures. It not only leads to nonlinear contact stiffness but also causes nonlinear damping. In some structures, the damping induced by frictional energy dissipation can account for up to 90% of the total structural damping [16, 20, 22, 45]. Figure 1.4 shows a set of frequency response curves for a typical connected structure, demonstrating that the system frequency response characteristics are closely related to the excitation amplitude, specifically in two aspects: (i) as the excitation amplitude increases, the system resonance frequency shifts gradually from the high frequency to the lower frequency and (ii) the system response amplitude decreases initially. Additionally, when intermittent separation and contact occur in the normal direction of the contact surfaces, the system frequency response curve may exhibit a 'tilted' peak [46–48], i.e. unstable solutions and multiple solutions.

When a connected structure is subjected to long-term vibration loads, material damage, surface morphology changes, or residual deformation can occur on the joint surfaces, a failure mechanism known as fretting failure [50–54]. Fretting failure includes fretting fatigue and fretting wear. Generally, fretting fatigue corresponds to cases where the relative displacement between contact surfaces is small. In such cases, stress concentration occurs in local contact areas, which may lead to crack formation or even fracture on the contact surface [51, 55]. In contrast, when fretting wear occurs, the relative displacement at the interface is larger. Typically, fretting wear can be divided into three stages based on the degree of wear: running-in wear, steady wear, and severe wear.

In the running-in wear stage, the contact surface morphology changes rapidly, accompanied by the generation of wear particles, expansion of wear scars, and increased wear depth, with significant changes in contact parameters (tangential

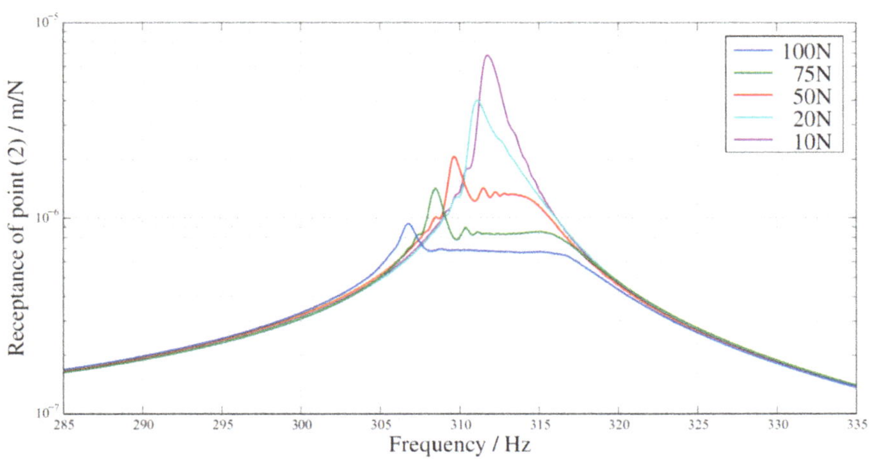


**Figure 1.4.** Frequency response curves of the Gaul resonator at different excitation amplitudes. (Reproduced with permission from [49]. Copyright 2018 Springer Nature.)

contact stiffness and friction coefficient). This stage is relatively short. In the steady wear stage, changes in the contact surface morphology stabilize, and a balance is achieved between the generation and removal of wear particles, with contact parameters remaining almost constant. This stage can last for a long time. As the degree of wear further increases, the contact surface reaches the severe wear stage, which can lead to mechanical failure and malfunction [56, 57].

Yoon *et al* [58] experimentally studied the fretting wear behavior of spherical contact under constant normal load and measured the evolution of the hysteresis loop. The results showed that the shape of the hysteresis loop changes with increasing fretting cycles—the amplitude of relative displacement gradually decreases, while the tangential friction force during the macroslip stage increases. Energy dissipation per cycle increases rapidly within the first 500 cycles and then stabilizes. Similar experimental results can also be found in [59, 60]. Fretting wear behavior not only causes structural failure but also alters the dynamic characteristics of joint structures due to changes in contact parameters [61–63], increasing the difficulty of controlling the structure.

## 1.4 Contact modeling of joint interfaces

Modeling the friction and wear behavior of connection interfaces can provide a basis for the design, optimization, control, life prediction, and reliability analysis of complex joint equipment. The contact between connection interfaces makes it impossible for existing technologies to directly observe the mechanical behavior of the contact surface without changing the contact conditions, including the distribution of contact pressure, the distribution of slip zone and sticking zone, and local deformation. This leads to a lack of unified understanding of the mechanical behavior of the contact surface, so it is unrealistic to expect a universal friction law that is applicable to the mechanical behavior of most connection interfaces [64].

Extensive experiments have shown that interface friction is related to many factors, such as the velocity, displacement, and condition of the contact surface of the contact pair. As a result, various contact friction models have been developed, such as the Coulomb friction model, the rate-dependent friction model, the static friction model, and the hysteresis friction model. These models have been reviewed in detail in the literature [65–70]. How to choose a suitable model depends on the specific application, such as the geometry of the contact body, the kinematic relationship generated by motion, the type of load, etc. This book divides the mechanical modeling methods of bolted connections into four categories: te spring element, thin layer element (TLE, also known as virtual material), zero-thickness element (ZTE), and friction hysteresis model.

### 1.4.1 Spring element

In the early stage, the mechanical behavior of bolted joints was always assumed to be linear, the connection stiffness thus was simulated using a single spring element [71] or a series of distributed springs with equal stiffness [72], as shown in figure 1.5(a). However, this modeling method fails to account for the impact of non-uniform

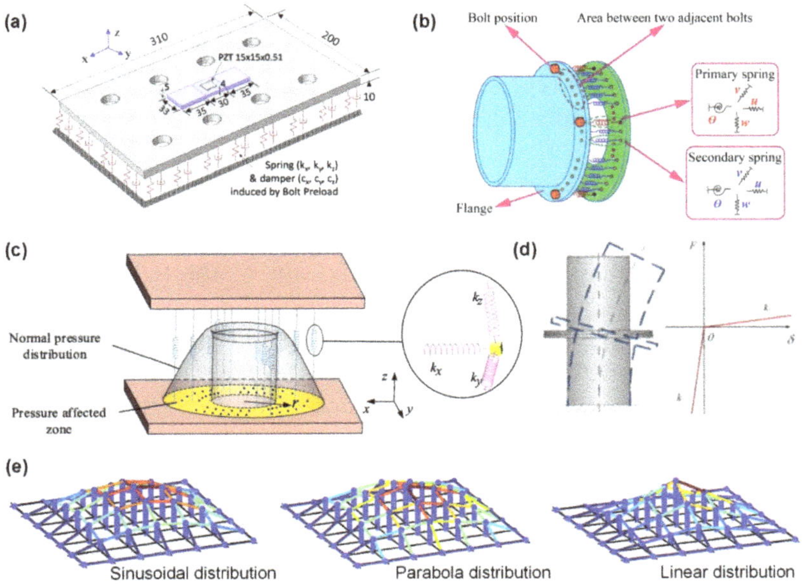


**Figure 1.5.** Modeling the joint by (a) a series of distributed springs with equal stiffness, (b) cosine-distributed spring stiffness to simulate the effect of the non-uniform contact pressure distribution, (c) a variable stiffness spring surface model, (d) a displacement-dependent nonlinear complex spring element, and (e) the axial stiffness of bolted joints in tension and compression. ((a) Reproduced from [72]. CC BY 4.0. (b) Reproduced with permission from [73]. Copyright 2024 Elsevier. (c) Reproduced with permission from [74]. Copyright 2022 Elsevier. (d) Reproduced with permission from [76]. Copyright 2012 Elsevier. (e) Reproduced with permission from [75]. Copyright 2022 Elsevier.)

stiffness distribution induced by the contact pressure distribution. Xing *et al* [73] used cosine-distributed spring elements to consider the effect of non-uniform contact pressure distribution in bolted connections, as shown in figure 1.5(b), in which they assumed that connection stiffness decreases linearly as the preload decays, thereby simulating the condition of inconsistent pre-tightening.

Liu *et al* [74] developed a spring surface model with variable stiffness where the spring stiffness is directly related to the contact pressure of joint interfaces, as shown in figure 1.5(c). The model parameters can be determined through optimization methods. Subsequently, they introduced a new modeling approach using displacement-dependent nonlinear complex spring elements with non-uniformly distributed parameters [75]. Sinusoidal, parabolic, and linearly distributed complex spring elements were used to simulate the non-uniform pressure distribution in the bolt influence zone, as shown in figure 1.5(e). The stiffness and damping displacement-dependent parameters of the complex spring elements were described by higher-order polynomials and estimated by an inverse identification method.

Considering the difference in axial stiffness of bolted connections during tension and compression, Luan *et al* [76] modeled the connection using a bilinear spring and analyzed the dynamic response of the bolted flange structure under coupled lateral and longitudinal vibrations, as shown in figure 1.5(d). To address the nonlinear friction at joint interfaces, Ahmadian *et al* [77] modeled the bolted connection using a combination of linear and nonlinear springs. A cubic stiffness and a viscous damping term were used to represent the saturation phenomena and energy loss at the connection interface under high-level vibrations.

### 1.4.2 Thin layer element

The TLE method models the interface action between two contacting parts as a continuum with very small thickness. It is also one of the most commonly used equivalent models of joints. The thin layer element can be easily integrated with commercial finite element software and has the advantage of being applicable to complex assembled structures with millions of degrees of freedom.

Wang *et al* [78] provided an effective modeling approach based on thin layer element theory and related the equivalent parameters to the mechanical characteristics of bolted joints, including tension and compression stiffness, bending stiffness, shear stiffness, and torsion stiffness, as shown in figure 1.6(a). In order to achieve forward prediction, Tian *et al* [79] proposed an analytical method for dynamic modeling of joint interfaces. According to Hertzian contact and fractal theory, they derived analytical solutions for the elastic modulus, shear modulus, Poisson's ratio, and density of thin layer elements.

The above studies mainly focus on the modal frequencies and vibration shapes of the connected structures. To further consider the friction damping caused by the joint, Wang *et al* [80] used a genetic algorithm to optimize the mechanical properties of the thin layer elements (see figure 1.6(b)) and minimize the error between the simulated dynamic characteristics and the experimental results, including the modal frequencies and frequency response functions (FRFs).

Chu *et al* [81] modeled the bolted joint interface as a thin layer of nonlinear elastic–plastic material. The material properties of the thin layer were defined by a

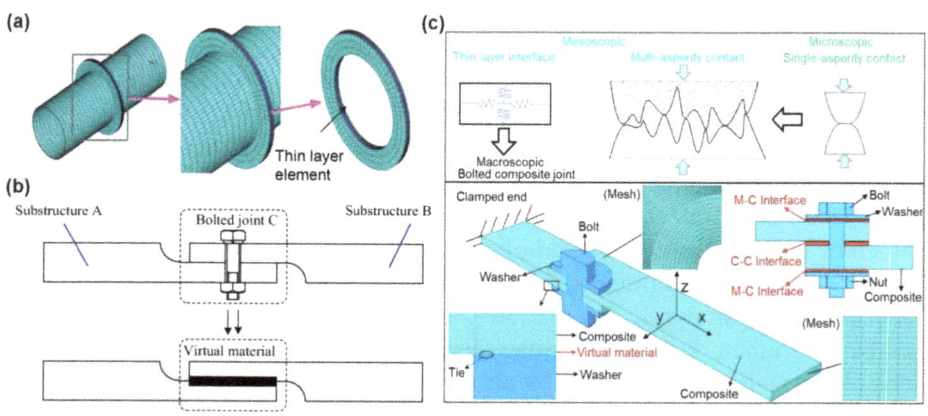


**Figure 1.6.** (a) Thin layer element, (b) modeling bolted joint interfaces as equivalent virtual materials, and (c) multi-scale model of a bolted composite joint with thin layer elements. ((a) Reproduced from [78]. CC BY 4.0. (b) Reproduced with permission from [80]. Copyright 2020 Springer Nature. (c) Reproduced with permission from [82]. Copyright 2019 Elsevier.)

bilinear elastic–plastic model and identified using a genetic algorithm. Zhang *et al* [82] used fractal theory to derive the complex contact moduli of an inhomogeneous joint interface characterized by multi-scale rough contact at the microscopic level, as shown in figure 1.6(c). These moduli were then applied to the properties of the thin layer element to describe the stiffness and damping of the connection interface.

### 1.4.3 Zero-thickness element

The concept of ZTEs was first introduced by Goodman *et al* [83] and further elaborated in detail in [84]. Figure 1.7(a) illustrates a schematic diagram of a ZTE with numbered nodes [85]. One major advantage of these elements is that they do not use the global contact search algorithm employed in master–slave contact models. The contact search algorithm is not necessary for modeling contact interfaces of joints, since the interfaces are always in contact and limited to small relative motions, which significantly saves computation time.

Mayer *et al* [85] developed a structural contact damping model based on the Masing model as the nonlinear constitutive law for ZETs. Gaul *et al* [86] improved this method by using a statistical approach based on Hertz and Mindlin contact models for individual spherical asperities to model the contact of rough surfaces. Actually, the surface roughness of each segment is not exactly the same.

Balaji *et al* [87] measured the surface topography using a high-resolution instrument and characterized the interface roughness on an element-by-element or segment-by-segment fashion, as shown in figure 1.7(b). They deduced a physics-based rough contact model and applied it to individual elements, allowing for the analysis of the effects of non-uniformly distributed surface roughness.

Porter *et al* [88] developed a new contact model that combines arbitrary normal and tangential displacement histories with a smooth tangential force–displacement relationship. The model considers both elastic and plastic normal contact of rough

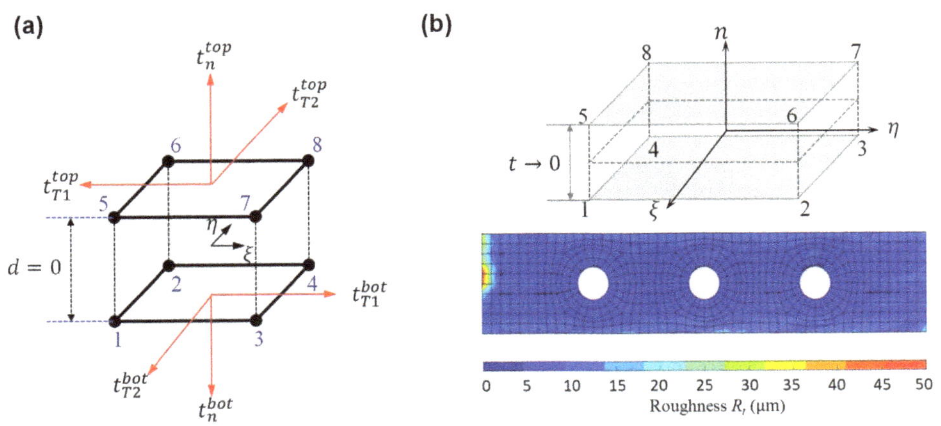


**Figure 1.7.** (a) Zero-thickness element and (b) zero-thickness element considering the measured multi-scale topography. ((a) Reproduced with permission from [85]. Copyright 2007 Elsevier. (b) Reproduced with permission from [87]. Copyright 2020 Elsevier.)

bodies, allowing for the simulation of the effects of incorporating plasticity into rough contact models. Additionally, this model accounts for the multi-scale characteristics of joint surfaces. Specifically, a gap function was used to describe the meso-scale topography and the rough contact model of each element, enabling blind prediction of the dynamic characteristics of assembled structures.

### 1.4.4 Friction hysteresis model

Generally, two types of interface modeling strategies can be used to reproduce tangential friction hysteresis at bolted joint interfaces: the distributed element model [48, 89, 90] and the whole interface model [20]. The distributed element model describes the frictional hysteresis behavior of local contact zones, while the characterization of the overall interface behavior is achieved through the discretization of the whole contact area, with the distributed element model simulating the frictional behavior of each discretized contact segment.

The most commonly used distributed element model is the Jenkins element [91], which consists of a linear spring and a Coulomb friction slider in series and can reproduce the tangential stick-slip behavior at joint interfaces. Yang *et al* [92, 93] introduced a linear spring into the Jenkins element to consider the effect of variable normal loads on friction behavior. However, these two models can not capture the coupled motion in two orthogonal directions on joint surfaces. To address this limitation, in [94–96] a Jenkins element is added in the orthogonal direction on the contact surface based on the above models, as shown in figure 1.8. This approach can be easily integrated into high-fidelity finite element models to analyze the dynamic response of joint structures and has been widely applied in the dynamic modeling of aircraft engine components [97]. Schwingshackl *et al* [89] introduced this method in the dynamic analysis of bolted flange joints. The authors of [98, 99] developed a multi-scale approach based on node-to-node contact elements to study the effect of roughness and fretting wear on the nonlinear dynamics of bolted joints, as shown in figure 1.8.

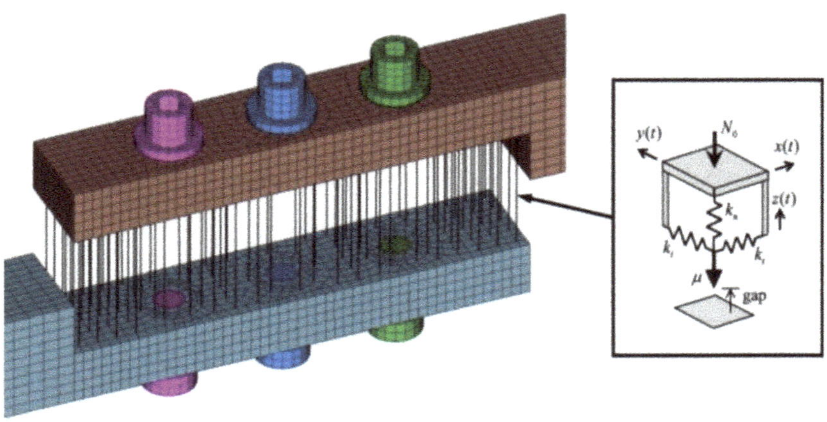


**Figure 1.8.** Node-to-node joint element. (Reproduced with permission from [99]. Copyright 2019 Elsevier.)

The advantage of the distributed element method is that it can accurately describe local phenomena at the contact segment. However, since it requires many node-to-node contact elements for the whole interface, this results in considerable nonlinear degrees of freedom in the model, thereby increasing computational costs. While this modeling approach is precise, computational efficiency becomes a major limiting factor when applied to complex structures.

The whole interface modeling strategy only requires one contact model to characterize the contact behavior of the interface, which greatly reduces the nonlinear degrees of freedom of joint structures. This type of model mainly includes the Menq shear layer model [100, 101], LuGre model [102] (see figure 1.9(a)), Valanis model [103–105] (see figure 1.9(b)), Iwan model [106, 107] (see figure 1.9(c)), Bouc–Wen model [108, 109], etc.

Gaul *et al* [18] were the first to apply the Valanis model, which is used to characterize the elastoplastic behavior of materials, to the hysteresis friction modeling of bolted joint interfaces. They integrated it as a nonlinear substructure module with a finite element program to analyze the dynamic response of a two-dimensional model of a space truss structure. The constitutive equation of the Valanis model is in differential form, and under certain parameter selections, it can degenerate into a bilinear model.

The Iwan model [106] was first used to simulate the elastoplastic behavior of materials. It consists of a series of Jenkins elements combined in series and parallel, as shown in figure 1.9(c). The constitutive relationship of the model can be obtained based on the distribution density function of critical sliding force (hereafter referred to as the Iwan density function) and classification statistical methods. Over time, many improved Iwan-type models [110–116] have been derived from the original Iwan model, which can be categorized into two types based on different modeling approaches.

The first type of modeling approach retains the essence of the Iwan model and seeks to improve the model accuracy and completeness through enhancements. Song *et al* [110] connected a linear spring in parallel to the original Iwan model to describe the residual stiffness phenomenon of bolted joint interfaces during the gross slip stage. Segalman *et al* [111], based on the power-law relationship between the excitation amplitude and the energy dissipation per cycle, proposed a four-parameter Iwan model by assuming the Iwan density function as a truncated

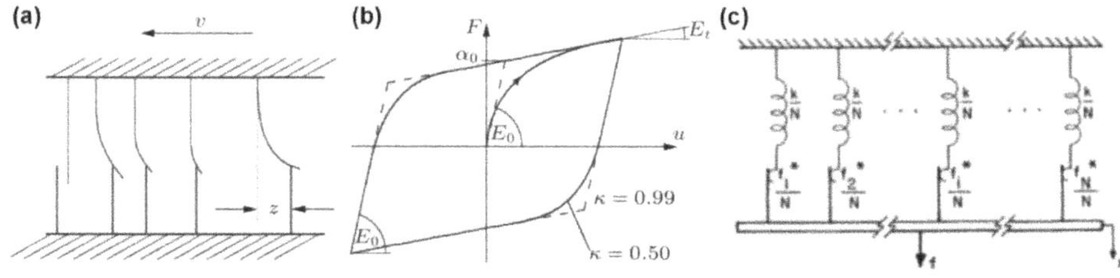


**Figure 1.9.** Representative whole interface model: (a) LuGre brush model, (b) Valanis model, and (c) Iwan model. (Reproduced from [69]. CC BY 4.0.)

power-law distribution with a single impulse. Wang *et al* [112], Li *et al* [113], Brake [114], Rajaei *et al* [115] and Ranjan *et al* [116] made a lot of improvements based on this model. The above-mentioned Iwan-type models artificially assume the Iwan density function, which lacks a clear physical meaning. However, the Iwan density function is, in fact, directly related to the contact pressure distribution. Li *et al* [117–119] derived the mapping relationship from contact pressure to the Iwan model density function. This idea has also been accepted and adopted in the literature [120–122].

The second modeling approach combines the Iwan model framework with the rough contact theory, which can be regarded as a multi-scale fusion model. Chen *et al* [123] simplified the tangential contact of asperities under a certain normal load into the Jenkins element in the parallel-series Iwan model, and then established a continuous function with clear physical meaning to describe the yield force distribution of the asperities on the joint interface, based on the rough contact theory and Coulomb friction law. Recently, Yang *et al* [124] considered the influence of surface roughness on the contact pressure distribution, modeled it as a twofold Weibull distribution, and derived the Iwan density function. Subsequently, they proposed a multi-scale friction hysteresis model based on multi-scale surface morphology decomposition and reconstruction to comprehensively consider the multi-scale morphology of the as-built surface and realize the forward prediction of friction hysteresis [125], as shown in figure 1.10.

## 1.5 Contribution and organization of this book

The fretting friction and wear in bolted connections is a key factor affecting structural performance and lifespan, especially under conditions of high load, high vibration, or cyclic impact. Therefore, studying fretting friction and wear phenomena is of great practical significance for improving the reliability of bolted connections and reducing maintenance frequency.

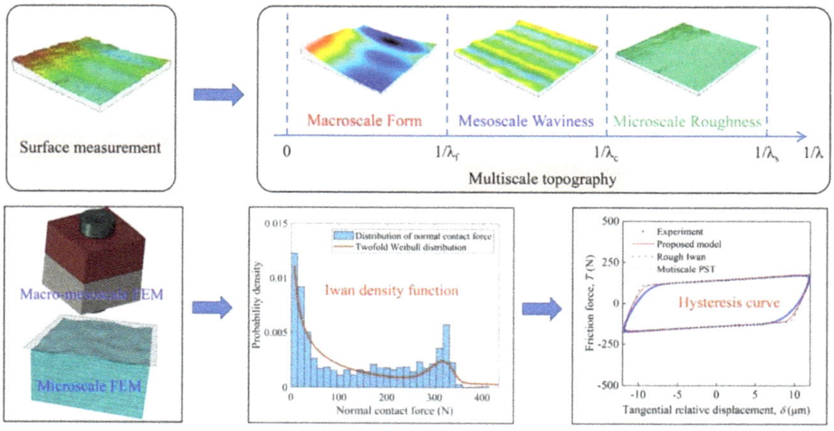


**Figure 1.10.** A multi-scale friction hysteresis model based on the framework of the Iwan model and fractal theory. (Reproduced with permission from [125]. Copyright 2024 Elsevier.)

This book summarizes the author's research in the past few years on fretting tests and mechanical modeling methods at bolted joint interfaces. The knowledge on bolt connection interface systematically introduced in this book has potential guiding significance for bolt connection structure design, bolt anti-loosening design, and health monitoring of jointed structures, etc.

This book involves many branches of mechanical engineering: contact mechanics, tribology, dynamics, experimental mechanics, modeling and numerical simulation. The rest of this book is organized as follows. Chapter 2 presents a fretting test apparatus to measure friction hysteresis at bolted joint interfaces. Some observed contact phenomena are explained theoretically through finite element analysis and lumped parameter models. Chapters 3 and 4 show the measured fretting wear behavior of bolted joint interfaces under sinusoidal and random oscillations, mainly including the evolution of contact parameters with increasing wear, the effect of surface roughness, the change in surface topography, and physics-based explanations. Chapter 5 introduces a generalized Iwan model and a detailed derivation of the Iwan density function. Chapter 6 applies the generalized Iwan model to bolted joint interfaces and further in dynamic response analysis of a three-degree-of-freedom joint system. Chapter 7 introduces a multi-scale contact model combining the multi-scale rough contact model with the discretized Iwan model. Finally, chapter 8 reconstructs the evolution equations of contact parameters and presents a fretting wear modeling method and application to the long-term dynamic analysis of jointed structures.

## Bibliography

[1] Galińska A 2020 Mechanical joining of fibre reinforced polymer composites to metals—a review. Part I: bolted joining *Polymers* **12** 2252

[2] Milan M T, Spinelli D, Bose Filho W W, Montezuma M F V and Tita V 2004 Failure analysis of a SAE 4340 steel locking bolt *Eng. Fail. Anal.* **11** 915–24

[3] American Airlines Flight 191 *Wikipedia* https://en.wikipedia.org/wiki/American_Airlines_Flight_191

[4] Porche 2014 Porsche to replace engines of current 911 GT3 models *Porche.com* https://porsche.com/usa/aboutporsche/pressreleases/pag/?id=E330392AE9044B5DC1257C9F005D2B11&pool=international-de

[5] Model X steering assist motor bolt recall *Tesla.com* https://tesla.com/en_ca/support/model-x-steering-assist-motor-bolt-recall

[6] Gregory J 2024 Boeing 737 Max 9: United Airlines finds loose bolts in jet inspections *BBC News Online* 9 Jan https://bbc.com/news/world-us-canada-67919436

[7] McCarthy M, Padhi G, Stanley W, McCarthy C and Lawlor V 2003 Bolted joints in composite aircraft structures *Composite Workshop on Joining and Assembling Technologies* p 11

[8] Croccolo D, Massimiliano D A, Stefano , Mattia M, Giorgio O, Chiara S and Muhammad H B T 2023 Failure of threaded connections: a literature review *Machines* **11** 212

[9] Sanliturk K Y, Ewins D J and Stanbridge A B 2001 Underplatform dampers for turbine blades: theoretical modeling, analysis, and comparison with experimental data *J. Eng. Gas Turbines Power—Trans. ASME* **123** 919–29

[10] Gola M M 2023 A general geometrical theory of turbine blade underplatform asymmetric dampers *Mech. Syst. Sig. Process.* **191** 110167
[11] Ferhatoglu E, Gastaldi C, Botto D and Zucca S 2022 An experimental and computational comparison of the dynamic response variability in a turbine blade with under-platform dampers *Mech. Syst. Sig. Process.* **172** 108987
[12] Botto D and Umer M 2018 A novel test rig to investigate under-platform damper dynamics *Mech. Syst. Sig. Process.* **100** 344–59
[13] Archard J F 1953 Contact and rubbing of flat surfaces *J. Appl. Phys.* **24** 981–8
[14] Fouvry S, Kapsa P and Vincent L 1996 Quantification of fretting damage *Wear* **200** 186–205
[15] Ungar E E 1964 Energy dissipation at structural joints: mechanisms and magnitudes *Technical Documentary Report* FDL-TDR-64-98 Research and Technology Division, Air Force Systems Command, United States Air Force
[16] Ungar E E 1973 The status of engineering knowledge concerning the damping of built-up structures *J. Sound Vib.* **26** 141–54
[17] Rogers P F and Boothroyd G 1975 Damping at metallic interfaces subjected to oscillating tangential loads *J. Eng. Ind.* **97** 1087–93
[18] Gaul L and Lenz J 1997 Nonlinear dynamics of structures assembled by bolted joints *Acta Mech.* **125** 169–81
[19] Süß D and Willner K 2015 Investigation of a jointed friction oscillator using the multiharmonic balance method *Mech. Syst. Sig. Process.* **52** 73–87
[20] Ames N M, Lauffer J P, Jew M D, Segalman D J, Gregory D L, Starr M J and Resor B R 2009 *Handbook on Dynamics of Jointed Structures* (Albuquerque, NM: Sandia National Laboratories) No. SAND2009-4164
[21] Sanati M, Terashima Y, Shamoto E and Park S S 2018 Development of a new method for joint damping identification in a bolted lap joint *J. Mech. Sci. Technol.* **32** 1975–83
[22] Padmanabhan K K and Murty A S R 1991 Damping in structural joints subjected to tangential loads *Proc. Inst. Mech. Eng.* C **205** 121–9
[23] Stanbridge A B, Ewins D J, Sanliturk K Y and Ferreira J V 2001 Experimental investigation of dry friction damping and cubic stiffness non-linearity *Int. Design Engineering Technical Conf. and Computers and Information in Engineering (9–12 September, Pittsburgh, PA)* (New York: The American Society of Mechanical Engineers) pp 2141–8
[24] Schwingshackl C W 2012 Measurement of friction contact parameters for nonlinear dynamic analysis *Topics in Modal Analysis I Proc. of the 30th IMAC, A Conf. on Structural Dynamics* vol 5 (New York: Springer) pp 167–77
[25] Filippi S, Akay A and Gola M M 2003 Measurement of tangential contact hysteresis during microslip *J. Tribol.* **126** 482–9
[26] Lavella M, Botto D and Gola M M 2013 Design of a high-precision, flat-on-flat fretting test apparatus with high temperature capability *Wear* **302** 1073–81
[27] Abad J, Franco J M, Celorrio R and Lezaun L 2012 Design of experiments and energy dissipation analysis for a contact mechanics 3D model of frictional bolted lap joints *Adv. Eng. Software* **45** 42–53
[28] Eriten M, Polycarpou A A and Bergman L A 2011 Development of a lap joint fretting apparatus *Exp. Mech.* **51** 1405–19

[29] Eriten M, Polycarpou A A and Bergman L A 2011 Effects of surface roughness and lubrication on the early stages of fretting of mechanical lap joints *Wear* **271** 2928–39

[30] Li D, Xu C, Botto D, Zhang Z and Gola M M 2020 A fretting test apparatus for measuring friction hysteresis of bolted joints *Tribol. Int.* **151** 106431

[31] Li D, Botto D, Xu C and Gola M M 2020 Fretting wear of bolted joint interfaces *Wear* **458** 203411

[32] Li D, Xu C, Li R and Zhang W 2022 Contact parameters evolution of bolted joint interface under transversal random vibrations *Wear* **500** 204351

[33] Li D, Botto D, Li R, Xu C and Zhang W 2022 Experimental and theoretical studies on friction contact of bolted joint interfaces *Int. J. Mech. Sci.* **236** 107773

[34] Ouyang H, Oldfield M J and Mottershead J E 2006 Experimental and theoretical studies of a bolted joint excited by a torsional dynamic load *Int. J. Mech. Sci.* **48** 1447–55

[35] Liu J, Ouyang H, Feng Z, Cai Z, Mo J, Peng J and Zhu M 2019 Dynamic behaviour of a bolted joint subjected to torsional excitation *Tribol. Int.* **140** 105877

[36] Wang Y, Ma Y, Hong J, Battiato G and Firrone C M 2024 Experimental studies on the energy dissipation of bolted structures with frictional interfaces: a review *Friction* **12** 1623–54

[37] Bickford J H 2007 *Introduction to the Design and Behavior of Bolted Joints: Non-gasketed Joints* (Boca Raton, FL: CRC Press)

[38] Jhang K Y, Quan H H, Ha J and Kim N Y 2006 Estimation of clamping force in high-tension bolts through ultrasonic velocity measurement *Ultrasonics* **44** e1339–42

[39] Mindlin R D 1949 Compliance of elastic bodies in contact *J. Appl. Mech.* **16** 259–68

[40] Mindlin R D, Mason W P, Osmer T J and Deresiewicz H 1952 Effects of an oscillating tangential force on the contact surfaces of elastic spheres *Proc. First US Natl Cong. Appl. Mech.* **1951** 203–8

[41] Johnson K L 1955 Surface interaction between elastically loaded bodies under tangential forces *Proc. R. Soc. London* A **230** 531–48

[42] Lubrication and Wear GroupHaines D J and Ollerton E 1963 Contact stress distributions on elliptical contact surfaces subjected to radial and tangential forces *Proc. Inst. Mech. Eng.* **177** 95–114

[43] Schwingshackl C W, Petrov E P and Ewins D J 2012 Measured and estimated friction interface parameters in a nonlinear dynamic analysis *Mech. Syst. Sig. Process.* **28** 574–84

[44] Fantetti A, Botto D, Zucca S and Schwingshackl C 2024 Guidelines to use input contact parameters for nonlinear dynamic analysis of jointed structures: results of a round robin test *Tribol. Int.* **191** 109158

[45] De Benedetti M, Garofalo G, Zumpano M and Barboni R 2007 On the damping effect due to bolted junctions in space structures subjected to pyro-shock *Acta Astronaut.* **60** 947–56

[46] Firrone C M and Zucca S 2011 Modelling friction contacts in structural dynamics and its application to turbine bladed disks *Numer. Anal.-Theory Appl.* **14** 301–34

[47] Zucca S, Firrone C M and Gola M 2013 Modeling underplatform dampers for turbine blades: a refined approach in the frequency domain *J. Vib. Control* **19** 1087–102

[48] Wu Y G, Li L, Fan Y, Zucca S, Gastaldi C and Ma H Y 2020 Design of dry friction and piezoelectric hybrid ring dampers for integrally bladed disks based on complex nonlinear modes *Comp. Struct.* **233** 106237

[49] Süß D, Janeba A and Willner K 2018 The Gaul resonator: experiments for the isolated investigation of a bolted lap joint *The Mechanics of Jointed Structures: Recent Research and*

*Open Challenges for Developing Predictive Models for Structural Dynamics* (Cham: Springer) pp 59–72
[50] Suh N P and Sin H C 1981 The genesis of friction *Wear* **69** 91–114
[51] Fouvry S, Duo P and Perruchaut P 2004 A quantitative approach of Ti–6Al–4V fretting damage: friction, wear and crack nucleation *Wear* **257** 916–29
[52] Elleuch K and Fouvry S 2005 Experimental and modelling aspects of abrasive wear of a A357 aluminium alloy under gross slip fretting conditions *Wear* **258** 40–9
[53] Ghosh A, Leonard B and Sadeghi F 2013 A stress based damage mechanics model to simulate fretting wear of Hertzian line contact in partial slip *Wear* **307** 87–99
[54] Li Z *et al* 2018 Fretting wear damage mechanism of uranium under various atmosphere and vacuum conditions *Materials* **11** 607
[55] De Pannemaecker A, Fouvry S, Brochu M and Buffiere J Y 2016 Identification of the fatigue stress intensity factor threshold for different load ratios *R*: from fretting fatigue to C (T) fatigue experiments *Int. J. Fatigue* **82** 211–25
[56] Zmitrowicz A 2006 Wear patterns and laws of wear—a review *J. Theor. Appl. Mech.* **44** 219–53
[57] Deuis R L, Subramanian C and Yellup J M 1997 Dry sliding wear of aluminium composites—a review *Compos. Sci. Technol.* **57** 415–35
[58] Yoon Y, Etsion I and Talke F E 2011 The evolution of fretting wear in a micro-spherical contact *Wear* **270** 567–75
[59] Liskiewicz T and Fouvry S 2005 Development of a friction energy capacity approach to predict the surface coating endurance under complex oscillating sliding conditions *Tribol. Int.* **38** 69–79
[60] Hirsch M R and Neu R W 2013 A simple model for friction evolution in fretting *Wear* **301** 517–23
[61] Fantetti A, Tamatam L R, Volvert M, Lawal I, Liu L, Salles L and Nowell D 2019 The impact of fretting wear on structural dynamics: experiment and simulation *Tribol. Int.* **138** 111–24
[62] Tamatam L R, Botto D and Zucca S 2023 A coupled approach to model wear effect on shrouded bladed disk dynamics *Int. J. Mech. Sci.* **237** 107816
[63] Lemoine E, Nélias D, Thouverez F and Vincent C 2020 Influence of fretting wear on bladed disks dynamic analysis *Tribol. Int.* **145** 106148
[64] Brake M R W 2018 *The Mechanics of Jointed Structures: Recent Research and Open Challenges for Developing Predictive Models for Structural Dynamics* (Berlin: Springer)
[65] Gaul L and Nitsche R 2001 The role of friction in mechanical joints *Appl. Mech. Rev.* **54** 93–106
[66] Ferri A A 1995 Friction damping and isolation systems *J. Mech. Des.* **117** 196–206
[67] Berger E 2002 Friction modeling for dynamic system simulation *Appl. Mech. Rev.* **55** 25–32
[68] Mathis A T, Balaji N N, Kuether R J, Brink A R, Brake M R and Quinn D D 2020 A review of damping models for structures with mechanical joints *Appl. Mech. Rev.* **72** 040802
[69] Liu Y, Zhu M, Lu X, Wang S and Li Z 2024 A review of cross-scale theoretical contact models for bolted joints interfaces *Coatings* **14** 539
[70] Shen M M and Yang X D 2024 Modeling of joint structure interface friction mechanics: a review *Adv. Mech.* **54** 477–521
[71] Nijgh M P 2016 *Loss of Preload in Pretensioned Bolts* (Delft: Delft University of Technology)

[72] Huynh T C, Dang N L and Kim J T 2018 Preload monitoring in bolted connection using piezoelectric-based smart interface *Sensors* **18** 2766
[73] Xing W C and Wang Y Q 2024 Dynamic modeling and vibration analysis of bolted flange joint disk-drum structures: theory and experiment *Int. J. Mech. Sci.* **272** 109186
[74] Liu H, Sun W, Du D and Liu X 2022 Modeling and free vibration analysis for bolted composite plate under inconsistent pre-tightening condition *Compos. Struct.* **292** 115634
[75] Liu X, Sun W, Liu H, Du D and Ma H 2022 Nonlinear vibration modeling and analysis of bolted thin plate based on non-uniformly distributed complex spring elements *J. Sound Vib.* **527** 116883
[76] Luan Y, Guan Z Q, Cheng G D and Liu S 2012 A simplified nonlinear dynamic model for the analysis of pipe structures with bolted flange joints *J. Sound Vib.* **331** 325–44
[77] Ahmadian H and Jalali H 2007 Identification of bolted lap joints parameters in assembled structures *Mech. Syst. Sig. Process.* **21** 1041–50
[78] Wang Z, Fei C W and Wang J J 2017 Equivalent simulation of mechanical characteristics for parametric modeling of bolted joint structures *Adv. Mech. Eng.* **9** 1687814017704360
[79] Tian H, Li B, Liu H, Mao K, Peng F and Huang X 2011 A new method of virtual material hypothesis-based dynamic modeling on fixed joint interface in machine tools *Int. J. Mach. Tools Manuf.* **51** 239–49
[80] Wang D and Fan X 2020 Nonlinear dynamic modeling for joint interfaces by combining equivalent linear mechanics with multi-objective optimization *Acta Mech. Solida Sin.* **33** 564–78
[81] Chu Y, Wen H and Chen T 2016 Nonlinear modeling and identification of an aluminum honeycomb panel with multiple bolts *Shock Vib.* **2016** 1276753
[82] Zhang Z, Xiao Y, Xie Y and Su Z 2019 Effects of contact between rough surfaces on the dynamic responses of bolted composite joints: multiscale modeling and numerical simulation *Compos. Struct.* **211** 13–23
[83] Goodman R E, Taylor R L and Brekke T L 1968 A model for the mechanics of jointed rock *J. Soil Mech. Found. Div.* **94** 637–59
[84] Hohberg J M 1992 *A Joint Element for the Nonlinear Dynamic Analysis of Arch Dams* (Basel: Birkhäuser)
[85] Mayer M H and Gaul L 2007 Segment-to-segment contact elements for modelling joint interfaces in finite element analysis *Mech. Syst. Sig. Process.* **21** 724–34
[86] Gaul L and Mayer M 2008 Modeling of contact interfaces in built-up structures by zero-thickness elements *Conf. Proc. IMAC 26: Conf. and Expo. on Structural Dynamics (Orlando, FL, 4–7 February)* (Bethel, CT: Society for Experimental Mechanics)
[87] Balaji N N, Chen W and Brake M R W 2020 Traction-based multi-scale nonlinear dynamic modeling of bolted joints: formulation, application, and trends in micro-scale interface evolution *Mech. Syst. Sig. Process.* **139** 106615
[88] Porter J H and Brake M R W 2023 Towards a predictive, physics-based friction model for the dynamics of jointed structures *Mech. Syst. Sig. Process.* **192** 110210
[89] Schwingshackl C W, Di Maio D, Sever I and Green J S 2013 Modeling and validation of the nonlinear dynamic behavior of bolted flange joints *J. Eng. Gas Turbines Power* **135** 122504
[90] Firrone C M, Zucca S and Gola M M 2011 The effect of underplatform dampers on the forced response of bladed disks by a coupled static/dynamic harmonic balance method *Int. J. Non Linear Mech.* **46** 363–75

[91] Jenkins G M 1962 Analysis of the stress–strain relationships in reactor grade graphite *Br. J. Appl. Phys.* **13** 30
[92] Yang B D and Menq C H 1998 Characterization of contact kinematics and application to the design of wedge dampers in turbomachinery blading: part 2—prediction of forced response and experimental verification *J. Eng. Gas Turbines Power-Trans. ASME* **120** 418–23
[93] Yang B D, Chu M L and Menq C H 1998 Stick–slip–separation analysis and non-linear stiffness and damping characterization of friction contacts having variable normal load *J. Sound Vib.* **210** 461–81
[94] Menq C H and Yang B D 1998 Non-linear spring resistance and friction damping of frictional constraint having two-dimensional motion *J. Sound Vib.* **217** 127–43
[95] Sanliturk K Y and Ewins D J 1996 Modelling two-dimensional friction contact and its application using harmonic balance method *J. Sound Vib.* **193** 511–23
[96] Gu W and Xu Z 2010 3D numerical friction contact model and its application to nonlinear blade damping *Turbo Expo: Power Land, Sea, Air (Glasgow, 14–18 June)* (New York: American Society of Mechanical Engineers) pp 809–17
[97] Yuan J, Chiara G, Enora D G and Benjamin C 2024 Friction damping for turbomachinery: a comprehensive review of modelling, design strategies, and testing capabilities *Prog. Aerosp. Sci.* **147** 101018
[98] Armand J, Salles L, Schwingshackl C W, Süss D and Willner K 2018 On the effects of roughness on the nonlinear dynamics of a bolted joint: a multiscale analysis *Eur. J. Mech.* A **70** 44–57
[99] Lacayo R, Pesaresi L, Groß J, Fochler D, Armand J, Salles L and Brake M 2019 Nonlinear modeling of structures with bolted joints: a comparison of two approaches based on a time-domain and frequency-domain solver *Mech. Syst. Sig. Process.* **114** 413–38
[100] Menq C H, Bielak J and Griffin J H 1986 The influence of microslip on vibratory response, part I: a new microslip model *J. Sound Vib.* **107** 279–93
[101] Menq C H, Griffin J H and Bielak J 1986 The influence of microslip on vibratory response. Part II: a comparison with experimental results *J. Sound Vib.* **107** 295–307
[102] De Wit C C, Olsson H, Astrom K J and Lischinsky P 1995 A new model for control of systems with friction *IEEE Trans. Autom. Control* **40** 419–25
[103] Zhang G and Liu F 2023 Equivalent dynamic modeling for the relative rotation of bolted joint interface using valanis model of hysteresis *Machines* **11** 342
[104] Abad J, Medel F J and Franco J M 2014 Determination of Valanis model parameters in a bolted lap joint: experimental and numerical analyses of frictional dissipation *Int. J. Mech. Sci.* **89** 289–98
[105] Jalali H, Jamia N, Friswell M I, Khodaparast H H and Taghipour J 2022 A generalization of the Valanis model for friction modelling *Mech. Syst. Sig. Process.* **179** 109339
[106] Iwan W D 1966 A distributed-element model for hysteresis and its steady-state dynamic response *ASME J. Appl. Mec.* **33** 893–900
[107] Iwan W D 1967 On a class of models for the yielding behavior of continuous and composite systems *ASME J. Appl. Mech.* **34** 612–7
[108] Charalampakis A E and Koumousis V K 2008 On the response and dissipated energy of Bouc–Wen hysteretic model *J. Sound Vib.* **309** 887–95
[109] Triantafyllou S P and Koumousis V K 2012 Bouc–Wen type hysteretic plane stress element *J. Eng. Mech.* **138** 235–46

[110] Song Y, Hartwigsen C J, McFarland D M, Vakakis A F and Bergman L A 2004 Simulation of dynamics of beam structures with bolted joints using adjusted Iwan beam elements *J. Sound Vib.* **273** 249–76
[111] Segalman D J 2002 A four-parameter Iwan model for lap-type joints *J. Appl. Mech.* **72** 752–60
[112] Wang D, Xu C, Fan X and Wan Q 2018 Reduced-order modeling approach for frictional stick-slip behaviors of joint interface *Mech. Syst. Sig. Process.* **103** 131–8
[113] Li Y and Hao Z 2016 A six-parameter Iwan model and its application *Mech. Syst. Sig. Process.* **68** 354–65
[114] Brake M R W 2017 A reduced Iwan model that includes pinning for bolted joint mechanics *Nonlinear Dyn.* **87** 1335–49
[115] Rajaei M and Ahmadian H 2014 Development of generalized Iwan model to simulate frictional contacts with variable normal loads *Appl. Math. Modell.* **38** 4006–18
[116] Ranjan P and Pandey A K 2021 Modeling of pinning phenomenon in Iwan model for bolted joint *Tribol. Int.* **161** 107071
[117] Li D, Botto D, Xu C, Liu T and Gola M 2019 A micro-slip friction modeling approach and its application in underplatform damper kinematics *Int. J. Mech. Sci.* **161** 105029
[118] Li D, Botto D, Xu C and Gola M 2020 A new approach for the determination of the Iwan density function in modeling friction contact *Int. J. Mech. Sci.* **180** 105671
[119] Li D, Xu C, Kang J and Zhang Z 2020 Modeling tangential friction based on contact pressure distribution for predicting dynamic responses of bolted joint structures *Nonlinear Dyn.* **101** 255–69
[120] Li C, Jiang Y, Qiao R and Miao X 2021 Modeling and parameters identification of the connection interface of bolted joints based on an improved micro-slip model *Mech. Syst. Sig. Process.* **153** 107514
[121] Zhao B, Wu F, Sun K, Mu X, Zhang Y and Sun Q 2021 Study on tangential stiffness nonlinear softening of bolted joint in friction-sliding process *Tribol. Int.* **156** 106856
[122] Li C, Miao X, Qiao R and Tang Q 2021 Modeling method of bolted joints with micro-slip features and its application in flanged cylindrical shell *Thin-Walled Struct.* **164** 107854
[123] Chen J, Zhang J, Hong J and Zhu L 2019 Modeling tangential contact of lap joints considering surface topography based on Iwan model *Tribol. Int.* **137** 66–75
[124] Yang H, Xu C and Guo N 2023 Modelling tangential friction considering contact pressure distribution of rough surfaces *Mech. Syst. Sig. Process.* **198** 110406
[125] Yang H, Li D, Sun J and Xu C 2024 Multiscale modeling of friction hysteresis at bolted joint interfaces *Int. J. Mech. Sci.* **282** 109586

# Part I

## Experiments

**Chao Xu and Dongwu Li**

# Chapter 2

# Fretting test method for bolted joint interface

Accurately measuring the frictional force, relative displacement, and bolt preload of bolted joint surfaces is a basic requirement for characterizing the nonlinear hysteresis behavior of the interface and extracting contact parameters. This chapter introduces a novel fretting test apparatus based on a piezoelectric actuator, which is exemplary for the development of fretting test rigs. The friction force and the bolt preload are measured using a load cell and a force washer, respectively. The interface relative motion is measured using a single laser vibrometer and a prime. Furthermore, the characteristics and measurement accuracy of the device are analyzed systematically.

## 2.1 Fretting test apparatus

This section describes a fretting test apparatus to experimentally investigate the friction contact behavior of bolted joints under transversal vibration [1]. The tangential friction force transmitted at joint interfaces, the relative displacement between joint interfaces, and the bolt preload are measured to evaluate the fretting behavior at joint interfaces. A detailed design scheme and measurement principles are provided.

### 2.1.1 Description of the test apparatus

Figure 2.1 shows a schematic diagram of the developed test rig and its main components. This set-up mainly consists of three parts: the excitation system, measurement system, and bolted joint specimen. The excitation is provided by a closed-loop controlled piezoelectric actuator, whose right end is connected to a leaf spring. This actuator can output stable vibration displacement and transmit it to the moving specimen through the leaf spring, causing relative motion between the fixed specimen and the moving specimen. The leaf spring is clamped between two C-shaped half-frames (the support in figure 2.1) by tightening two M12 bolts. The two half-frames are machined from a monolithic steel block and make up the

doi:10.1088/978-0-7503-6214-6ch2  

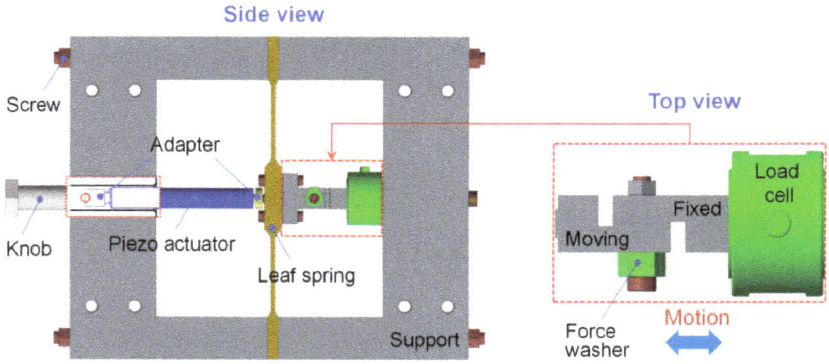


**Figure 2.1.** Sketch of the developed test apparatus and main components. (Adapted with permission from [1]. Copyright 2020 Elsevier.)

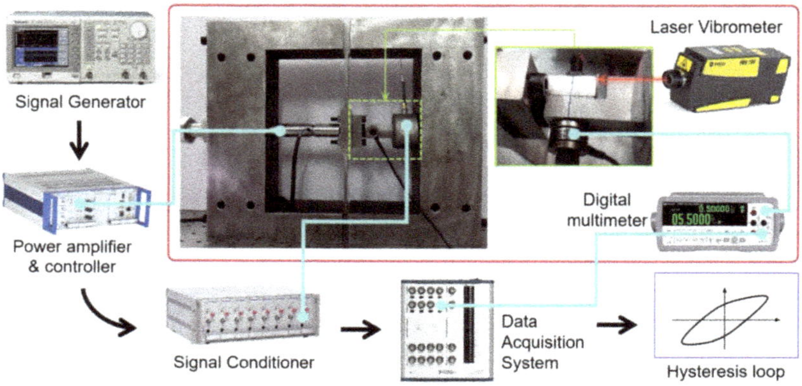


**Figure 2.2.** Photograph of the test apparatus and schematic diagram of signal acquisition.

mainframe of the rig. The mainframe forms an O-shaped closed gate so that all the internal forces are self-balanced. The moving and fixed specimens are tightened through an M6 bolt and in contact with each other. The bolt preload is measured using a force washer. The relative displacement between the fixed and moving specimens is measured using a laser vibrometer, as shown in figure 2.2. The right end of the fixed specimen is connected to a load cell through a thread (see the top view in figure 2.1), which is used to measure the friction force transmitted through the bolted joint interface. The contact surfaces and the axis of the load cell are carefully aligned so that the load cell measures the tangential contact force with great accuracy. The right end of the load cell is connected to the right support of the mainframe through an M10 screw.

The test apparatus is placed on an optical vibration isolation table to reduce the effects of external vibration environment on the measurements. When assembling the experimental set-up, the subsystems on both sides of the leaf spring are installed separately, and then connected by two M12 studs. Before and after each experiment, there is no need to disassemble the left excitation system, which greatly reduces the assembly time (no more than 20 min). Figure 2.2 shows a photograph of the experimental set-up and schematic diagram of signal transmission. In the following, we will provide a detailed introduction to the composition and working principles of each subsystem.

### 2.1.2 Excitation system

The excitation system consists of a piezoelectric actuator, a signal generator, a power amplifier, a leaf spring, a knob, two steel balls, and two adapters, as shown in figure 2.3. The piezoelectric actuator (PSt150/14/100VS20, Coremorrow Inc.) is preloaded, which outputs displacement excitation after receiving signals of specific frequencies and waveforms transmitted by the signal generator and power amplifier (E-505, Physik Instrument Inc.). The closed-loop control of the piezoelectric actuator is achieved through a servo controller (E-509.x1, Physik Instrument Inc.) and an integrated position feedback sensor (also called SGS sensor) to ensure stable output during fretting tests.

The actuator has nanoscale resolution and can output high-frequency excitation, but it cannot withstand radial shear loads that could lead to premature failure, therefore, an uncoupling system is devised. This uncoupling system consists of two adapters and two #10 steel balls, as shown in figure 2.3. The adapters are placed at both ends of the actuator and make contact with the balls, used to fully release all degrees of freedom of the piezoelectric actuator except for axial degrees of freedom during assembly and testing. If the actuator is displaced along the transverse direction, the contact angle between the sphere and the adapter groove changes and gives a counter-reaction. This counter-reaction, which is opposite to the displacement, is enough to keep the piezoelectric actuator in the right position if the transverse displacement is small. In addition, in order to ensure that the contact between the right steel ball and the leaf spring, and between the left steel ball and the knob does not separate, a preload is applied by screwing the knob during the initial installation of the device. The magnitude of this preload does not affect the measured results.

As mentioned before, due to the spherical contact at both ends of the actuator, it only can push the joint specimen, but cannot pull it back. That is why there is a leaf spring locating between the actuator and the moving specimen. The leaf spring (see figure 2.4) not only plays the role of pulling back the specimen after reaching the maximum position, but also ensures the stability of the test rig by limiting the

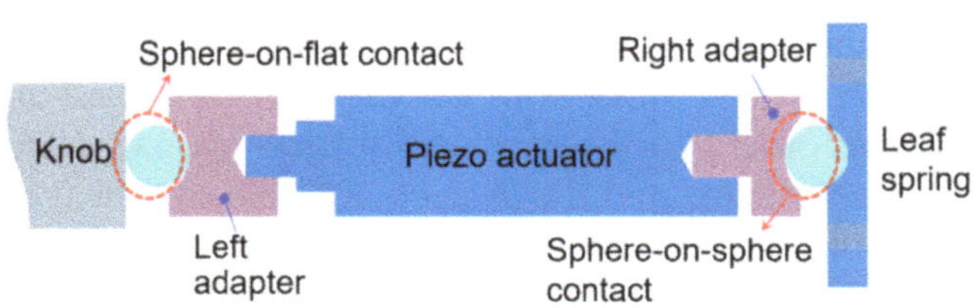


**Figure 2.3.** The main components of the excitation system and the installation diagram of the piezoelectric actuator. (Adapted with permission from [1]. Copyright 2020 Elsevier.)

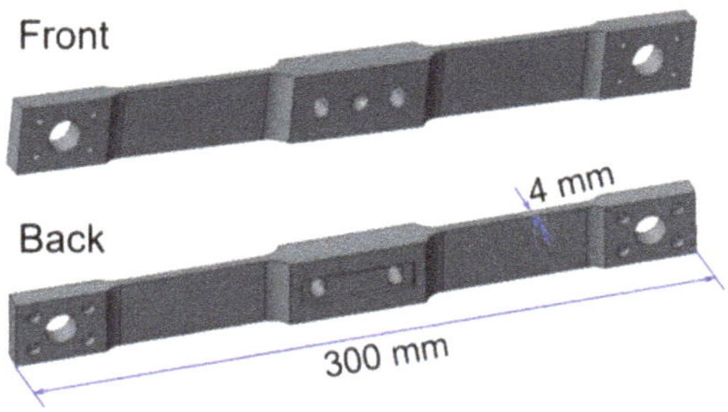


**Figure 2.4.** Schematic of the front and back sides of the leaf spring. (Adapted with permission from [1]. Copyright 2020 Elsevier.)

out-of-phase vibration. Without this leaf spring, any small lateral load could potentially cause the entire device to lose stability. Furthermore, to avoid resonance, the leaf spring is designed to have its natural frequencies higher than the operating frequency of the test rig. However, the stiffness of the leaf spring cannot be too large, otherwise the output capacity of the piezoelectric actuator will be limited in this situation. Therefore, the leaf spring was designed with a stiffness equal to 5% of the actuator stiffness (40 kN mm$^{-1}$) and with a first natural frequency of about 405 Hz. The resonance is far from the operating frequency (25 Hz) and the leaf spring is soft enough to allow the piezoelectric actuator to displace the moving specimens up to gross slip regime.

### 2.1.3 Measurement system

In this experiment, there are three variables that need to be measured, namely the tangential friction force transmitted through the joint interface, the bolt preload, and the relative displacement between the connected surfaces. These measured data are used to draw hysteresis curves that characterize the interface friction hysteresis behavior and to extract the tangential contact stiffness and friction coefficient of the interface.

The measurement of tangential friction force is achieved using a load cell (1061V2, Dytran Instruments Inc.), which has a measurement range of ±2224 N and a sensitivity of 2.25 mV N$^{-1}$.

The measurement accuracy of the bolt preload heavily determines the identification accuracy of the friction coefficient of the connected interfaces. At present, there are three main methods for measuring bolt preload, including the torque-controlled method, the strain gauge method, and the direct measurement method based on force washers. In [2, 3] the bolt preload was determined using the torque-controlled method according to the empirical torque–preload relationship. Given the torque measurement errors and the dispersion of the empirical coefficient in the torque–preload formula [4], the measured preload may be inaccurate. Li *et al* [5] employed a strain gauges attached to the bolt to measure the deformation of the bolt and thus obtain the corresponding preload. This method requires special processing of bolts to facilitate the placement of sensor wires. Abad *et al* [6] directly measured the preload using an annulus force washer in their test device. Compared to other methods, the direct measurement method is simple to operate and has high accuracy. Here, an annulus force washer (KMR/20 kN for M6 bolt, HBM Inc., 1.7 mV $V^{-1}$ sensitivity) is employed to directly monitor the bolt preload. A DC power supply (GPD-3303S, GW Instek Inc.) feeds the force washer with a 2.5 V DC voltage. A high-precision 5.5 digit digital multimeter (34450 A, Keysight Technologies Inc.) is used to record the output voltage of the force washer in real time.

Regarding the measurement of the interface relative displacement, non-contact measurement methods are always used. Eriten *et al* [2] measured the tangential displacement of joint specimens at the moving end using a single laser nano sensor and a small mirror, which assumed high rigidity at the fixed end and neglected the deformation of the fixed specimens. Schwingshackl *et al* [7] utilized a single laser Doppler vibrometer (LDV) to measure the displacement of the moving contact surface based on the assumption that the fixed contact surface has negligible movement. However, in fact, due to the deformation of the fixed-end support, the fixed-end surface will more or less undergo certain deformation. Therefore, there is an error in this measurement method, resulting in a smaller measured contact stiffness. Section 2.3 shows some measurements that detect the difference between the relative and absolute displacements. Nowadays, two LDVs are usually used to measure true relative displacement [8–11]. In this test rig, a novel measurement method is utilized, combining a single laser vibrometer (PDV-100, Polytec Inc.) with a prism without any approximation, as shown in figure 2.5. The laser vibrometer is placed on an optical isolation table to reduce the influence of environmental vibration on the measured results. The prism with a size of 10 mm × 10 mm × 10 mm is pasted on the fixed specimen to reflect the laser beam on the target measurement point of the moving specimen. The prism displacement represents the displacement of the fixed specimen. A polished lightweight aluminum alloy plate with a thickness of 0.5 mm is pasted on the end of the moving specimen as a mirror to reflect the laser beam, as shown in figure 2.5(a). The displacement of the plate represents the displacement of the moving specimen. Figure 2.5(b) shows the installation positions of the prism and the mirror. Additionally, it should be noted that the adhesion of the prism and mirror must be firm, reliable and easy to disassemble. As shown in figure 2.5(b), before each test, a layer of evenly distributed

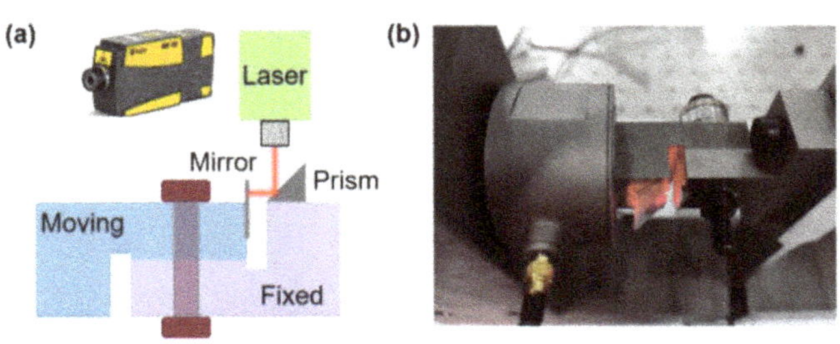


**Figure 2.5.** (a) Schematic of the measurement method of relative displacement between the fixed and moving specimens. (b) Installation location of the prism and the mirror. (Adapted with permission from [1]. Copyright 2020 Elsevier.)

paraffin wax is first used to paste the prism and the mirror on the end of the corresponding specimens. Then they are wrapped with hot melted glue to prevent them from vibration-induced movement.

Figure 2.6 illustrates the principle diagram of measuring relative displacement with only one laser head. The green dotted boxes represent the position of the moving and fixed specimens at the initial moment $t_1$. The blue and purple shaded areas indicate the positions of the moving and fixed specimens at the initial moment $t_2$, respectively. During the time interval $\Delta t = t_2 - t_1$, the laser spot changes from point A to A′, and the mirror position changes from point B to B′. The symbols $u_M$ and $u_F$ denote the displacements of the moving and fixed specimens, respectively. At moment $t_1$, the distance from the laser head to the mirror on the moving specimen is

$$d_{t_1} = |\mathrm{OA}'| + |\mathrm{A}'\mathrm{B}'|. \tag{2.1}$$

At moment $t_2$ the distance becomes

$$d_{t_2} = |\mathrm{OA}| + |\mathrm{AB}|, \tag{2.2}$$

where

$$|\mathrm{OA}| = |\mathrm{OA}'| + u_F \tan(\alpha),$$

$$|\mathrm{AB}| = |\mathrm{A}'\mathrm{B}'| - u_M, \tag{2.3}$$

and $\alpha$ represents the 45° angle of the prism. Substituting equation (2.3) into equation (2.2) yields the distance at moment $t_2$,

$$d_{t_2} = |\mathrm{OA}'| + |\mathrm{A}'\mathrm{B}'| + u_F - u_M. \tag{2.4}$$

Therefore, the displacement measured by the laser head within a time interval $\Delta t$ is

$$|d_{t_2} - d_{t_1}| = |u_F - u_M| = \delta. \tag{2.5}$$

Equation (2.5) demonstrates that the relative displacement $\delta$ can be measured using a single LDV and a prism with a inclined angle of 45°. The laser beam needs to be carefully adjusted before each test to ensure that a high-quality reflected laser beam is received. Moreover, the laser reflection point should be focused as close to the joint surface as possible.

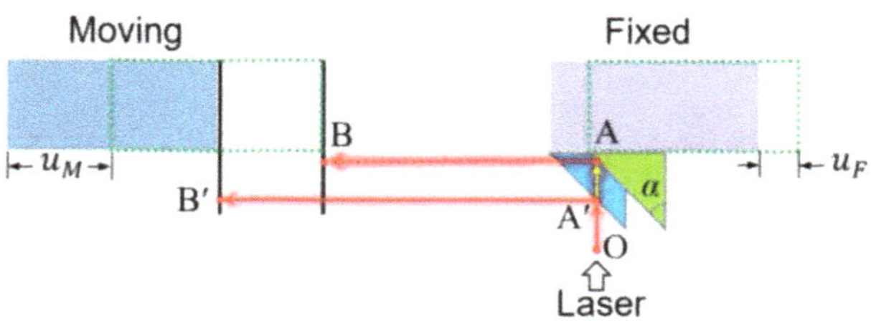


**Figure 2.6.** Principle of relative displacement measurement method using a single laser head. (Adapted with permission from [1]. Copyright 2020 Elsevier.)

The friction force and displacement signals are acquired using a 16 bit data acquisition board (PCI-6251, National Instruments Inc.) and an in-house LabView code. Before acquisition, signals are conditioned with a 12 channel signal conditioner (YE3826A, Sinocer Inc.). To monitor the stability of the actuator, the displacement of the actuator tip is also acquired with the built-in SGS sensor. The sampling frequency of both force and displacement signals is 5 kHz and the sampling number is 50 000.

### 2.1.4 Bolted joint specimens

The bolted joint specimens under tests were made of ASTM 304 stainless steel with a yield strength of 215 MPa. The tested bolts are M6, carbon steel, class 8.8. Figure 2.7(a) depicts the assembly diagram and the dimensions of the connected parts. The nominal contact area is about 350 mm$^2$. Before each test, the contact surfaces of the specimen were cleaned in an ultrasonic cleaner filled with alcohol for 30 min to avoid external particles and machine oil affecting the experimental results, and then dried in a constant temperature oven to ensure dry friction conditions. In addition, to reduce the influence of the joint specimen assembly error on the experimental results, a feeler gauge is used to ensure that the two through-holes are as coaxial as possible, as shown in figure 2.7(b).

The tested joint specimens were machined using wire cutting technology, so the roughness of the contact surfaces was high, which was not conducive to the formation of stable friction. In order to reduce the roughness of the contact surfaces and eliminate the influence of large-size wear particles on the test results as much as

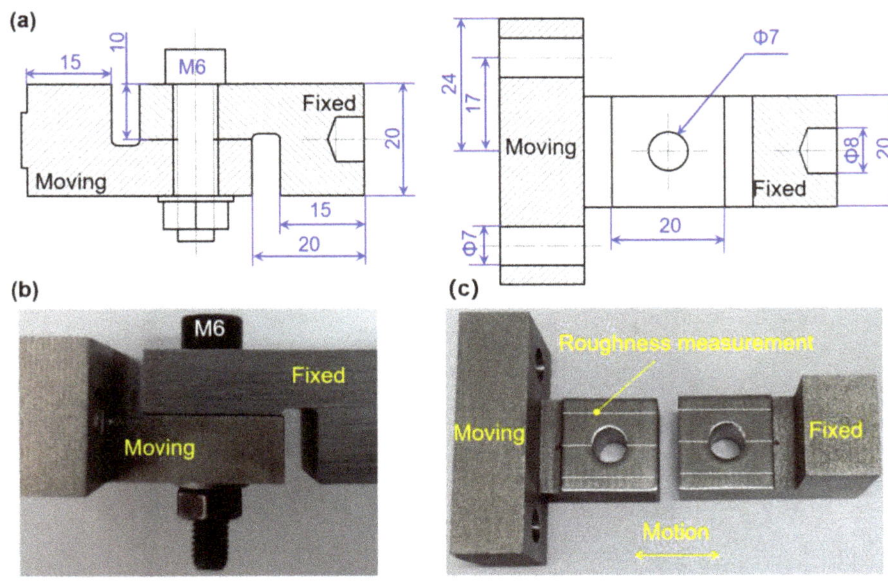


**Figure 2.7.** (a) Dimensions of the bolted joint specimen (the units are mm), (b) photograph of the joint specimen, and (c) roughness measurement path of the contact surfaces. (Adapted with permission from [1]. Copyright 2020 Elsevier.)

possible, two different grits of sandpaper (first grit 800 and then grit 1200) were used to carefully hand-polish the contact surfaces in sequence.

The surface roughness $R_a$ of the joint specimens was measured by a portable roughness profilometer (TR200, Jitai Keyi Inc.), and the white line segment in figure 2.7(c) represents the roughness measurement path. The test length along each long white line is 4 mm, and that along each short white line is 2.4 mm. The average value of the roughness of each line segment is regarded as the roughness of the contact surface. The roughness of the contact surfaces of the moving and fixed specimens in the tests are 0.78 $\mu$m and 0.91 $\mu$m, respectively. All tests were conducted at room temperature.

## 2.2 Performance analysis of the test apparatus

### 2.2.1 Typical experimental results

A typical test was preformed to illustrate the analysis method of the experimental results and the extraction method of contact parameters. In this case, the bolt preload is $N_0 = 900$ N, the nominal output displacement amplitude of the actuator tip is $\Delta x = 50$ $\mu$m, and the excitation frequency of the applied sine wave is $f = 25$ Hz. The nominal output displacement amplitude of the actuator is obtained according to the relationship between the piezoelectric equation and the input voltage, which differs slightly from the actual displacement value measured by the built-in SGS sensor. Figure 2.8 shows the time-domain curves of the measured interface relative displacement $\delta$, the actual output displacement of the actuator, and the tangential friction force $T$. Each curve contains 20 cycles, from which it can be seen that the experimental set-up is very stable during the testing process.

The measured friction force and the relative displacement form a hysteresis loop, which characterizes the nonlinear behavior of the joint interface, as shown in

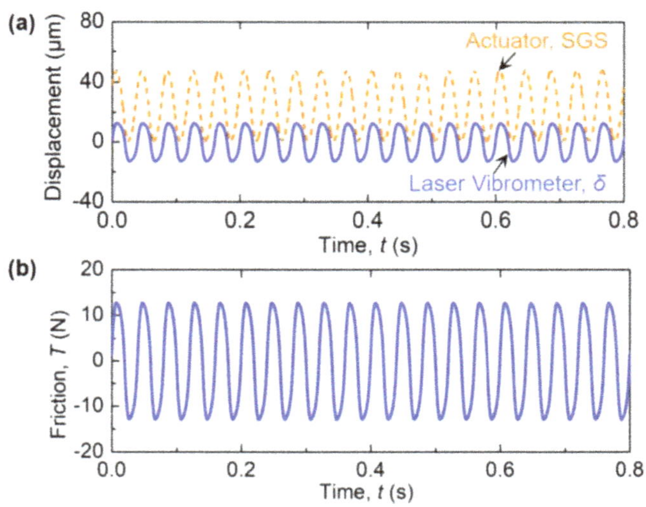


**Figure 2.8.** (a) Interface relative displacement $\delta$ measured by the laser vibrometer and the actuator tip displacement measured by the built-in SGS sensor and (b) measured friction force.

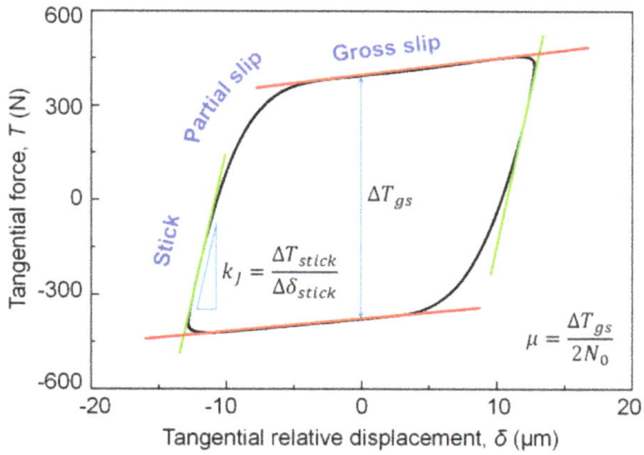


**Figure 2.9.** An exemplary hysteresis loop and schematic diagram of the contact parameter (including tangential contact stiffness and friction coefficient) extraction method. (Adapted with permission from [1]. Copyright 2020 Elsevier.)

figure 2.9. The area enclosed by the hysteresis loop represents the energy dissipation caused by interface friction per cycle. From the hysteresis curve, the motion state of the connected interface can be identified. Taking the loading stage as an example, the connected interface is first in a stick state, exhibiting linear elastic behavior. As the load increases, local sliding occurs between the interfaces, known as partial slip (also called microslip). Subsequently, the continuously increasing load causes the whole interface to slide, known as gross slip (also called macroslip). In some fretting tests, the friction force remains constant when the interface is in the gross slip state. Nevertheless, the measured friction force in this experiment remains variable during the gross slip stage and is approximately linearly correlated with relative displacement, that is, the 'residual stiffness' phenomenon. This phenomenon and its causes will be detailed in the next section.

On the premise that the preload force is known, the hysteresis loop can be characterized by the tangential contact stiffness and friction coefficient. The tangential contact stiffness is defined as the slope of the hysteresis curve in the stick regime, i.e. $k_J = \Delta T_{\text{stick}}/\Delta\delta_{\text{stick}}$, where the force increment $\Delta T_{\text{stick}}$ and displacement increment $\Delta\delta_{\text{stick}}$ are determined by choosing 120 consecutive points of the linear portion of the loop after the motion is reversed and performing a linear fit on them, as shown in figure 2.9. Essentially, the measured tangential stiffness $k_J$ is the overall joint stiffness that includes, in addition to the target contact interface, contributions of contact interfaces at the washer, the screw head, and the nut. The contact stiffness $k_t$ of the target surface can be inferred by $k_J$. A more detailed analysis of these stiffness contributions will be given in the next section.

The friction coefficient generally is defined as the ratio of the friction force during the gross slip regime to the normal load, but it does not apply to bolted joint interfaces due to changes in sliding friction force. Here, the friction coefficient can be estimated by the ratio of the distance between two red lines in figure 2.9 over twice the bolt preload, $\mu = \Delta T_{gs}/2N_0$.

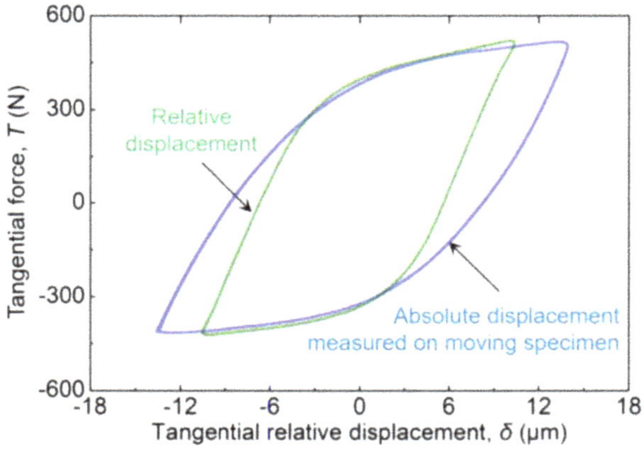


**Figure 2.10.** Comparison of measured hysteresis loops by different methods (considering the motion of the fixed specimen or not). (Adapted with permission from [1]. Copyright 2020 Elsevier.)

Figure 2.10 compares two measured hysteresis loops under the same operating scenario, plotted from the tangential force as a function of the absolute displacement (the blue line) and the relative displacement (the green line). The absolute displacement was measured by reflecting the laser beam onto the moving specimen using a prism fixed on the optical isolation table. It can be seen that the relative displacement amplitude is noticeably lower, about 40%, than the absolute displacement amplitude. Correspondingly, there are differences in the estimated contact stiffness and the contact stiffness based on the relative displacement is larger. Therefore, the displacement of the fixed specimen cannot be neglected in fretting tests.

### 2.2.2 Repeatability

Tests were conducted under the same preload and excitation to evaluate the repeatability of the measured hysteresis curves and estimated contact parameters. For this purpose, the same specimen pairs were assembled, tested, and disassembled three times. The measurement for each test lasted 1 s. Figure 2.11 shows the measured hysteresis loops under a sinusoidal excitation with 40 $\mu$m amplitude and 25 Hz frequency. Two bolt preloads ($N_0 = 235$ and 405 N) were considered. Each loops contain 25 cycles. During this period, the change in bolt preload is less than 0.5%. It can be seen from the degree of dispersion of hysteresis curves that the experimental results after different assemblies have good repeatability. This means the measurements are stable.

Table 2.1 lists the contact parameters (tangential contact stiffness and friction coefficient) estimated from the hysteresis loops, and the dissipated energy per cycle. The coefficient of variation was used to assess the dispersion of contact parameters and dissipated energy due to assembly uncertainty. The coefficient of variation is a statistical measure of the dispersion of data points in a data series around the mean and defined as the ratio of the standard deviation of measured data to its mean. The results show that the coefficients of variation for friction coefficient and energy dissipation are both less than 4%, while the coefficient of variation for contact

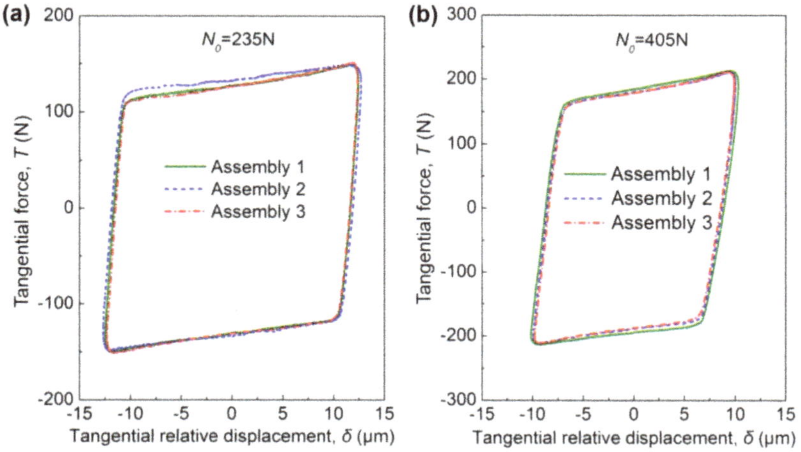


**Figure 2.11.** Measured hysteresis loops under three repeated assemblies and two bolt preloads: (a) $N_0 = 235$ N and (b) $N_0 = 405$ N. (Reproduced with permission from [1]. Copyright 2020 Elsevier.)

**Table 2.1.** Assessment of the repeatability of contact parameters and energy dissipation per cycle. (Reproduced with permission from [1]. Copyright 2020 Elsevier.)

| | Contact stiffness (N $\mu$m$^{-1}$) | | Friction coefficient | | Energy dissipation (mJ) | |
|---|---|---|---|---|---|---|
| Contact parameters | $N_0 = 235$ N | $N_0 = 405$ N | $N_0 = 235$ N | $N_0 = 405$ N | $N_0 = 235$ N | $N_0 = 405$ N |
| Assembly 1 | 155.3 | 155.1 | 0.549 | 0.467 | 6.06 | 6.73 |
| Assembly 2 | 143.6 | 166.2 | 0.566 | 0.455 | 6.32 | 6.40 |
| Assembly 3 | 158.4 | 171.3 | 0.548 | 0.453 | 5.99 | 6.29 |
| Coefficient of variation | 5.12% | 5.04% | 1.82% | 1.65% | 2.84% | 3.54% |

stiffness is slightly higher, but also below 5.5%. In view of this, the assembly uncertainty has little impact on the experimental results, and the repeatability of the test device is good. However, it should be noted that this is the result obtained after strictly following the experimental assembly steps. If the assembly steps are not followed, the dispersion of the experimental results may be significant.

### 2.2.3 Effect of excitation frequency

In the fretting test rig the controlled variable should be the relative displacement between the contact surfaces but this closed-loop control has not been implemented yet. Actually, the test rig here is semi-closed-loop controlled, in which only the actuator is closed-loop controlled. Therefore, the excitation frequency may affect the experimental results. Figure 2.12 shows the hysteresis curves measured under two

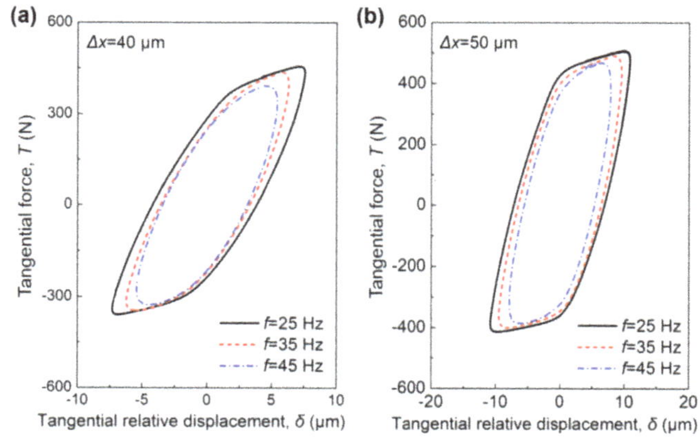


**Figure 2.12.** Measured hysteresis loops under different excitation frequencies and displacement excitation amplitudes: (a) $\Delta x = 40\ \mu m$ and (b) $\Delta x = 50\ \mu m$.

sets of excitation amplitudes and different excitation frequencies. It can be seen that even if the nominal displacement $\Delta x$ of the actuator is the same, the difference between each curve is noticeable. Under the same excitation amplitude, the lower the excitation frequency, the greater the relative displacement amplitude between the joint interfaces, and the correspondingly greater the energy dissipation.

The nominal displacement of the actuator is known as a function of the input voltage through a piezoelectric calibration curve. The true displacement of the piezoelectric tip was driven by the amplifier, whose average power depends on the operating frequency $f$,

$$P = U^2 fC, \tag{2.6}$$

where $U$ denotes the input voltage and $C$ the electrostatic capacity of the piezoelectric. For the same input voltage, the higher the operating frequency, the higher the average power. A higher operating frequency causes the actuator to heat up with the detriment of the actuator's performance. Figure 2.13 depicts the imposed displacement recorded by the internal SGS of the actuator. The results showed that the amplitude of the piezoelectric tip decreases as the operating frequency increases.

## 2.3 Measurement accuracy analysis

Different to the common fretting test rigs, the test rig introduced in this chapter does not allow a direct measurement of the frictional force at the target contact interface due to the special architecture of the bolted joint. In this experiment, the force measured by the load sensor not only includes the frictional force at the target interface, but also includes other contact forces at the bolt connection (contacts between the screw head, the washer, the nut, and the target interface) and the elastic force in the bolt shank due to bending deformation.

Therefore, this section establishes a finite element (FE) model and a simplified five degree of freedom lumped-parameter model for the bolted joint specimen to analyze

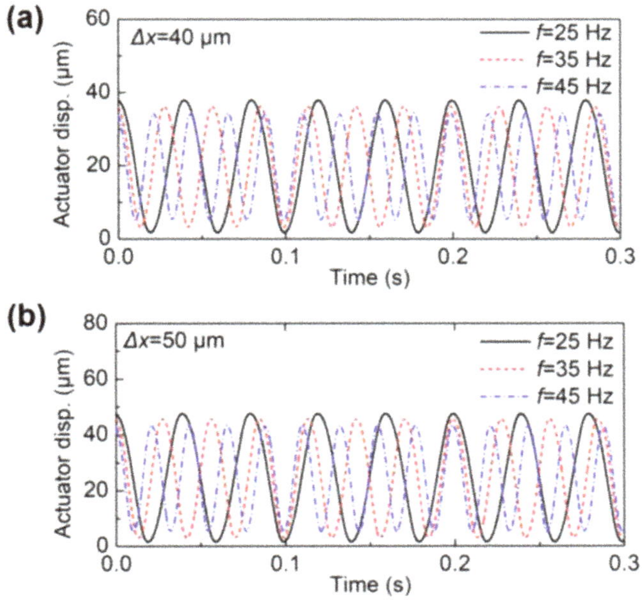


**Figure 2.13.** Actual output displacement of actuators under different excitation frequencies and displacement excitation amplitudes: (a) $\Delta x = 40\ \mu$m and (b) $\Delta x = 50\ \mu$m.

the force transfer characteristics of the local connect area. Based on the model analysis, a correction scheme for inferring the tangential contact stiffness at the target interface is given. The following analysis takes the case of $N_0 = 470$ N and $\Delta x = 40\ \mu$m as an example.

### 2.3.1 Finite element contact analysis

The FE software ABAQUS was used to model and analyze the contact of the bolted joint specimens, as shown in figure 2.14. All components in the model (including the bolt shank, nut, annular force washer, fixed specimens, and moving specimens) were meshed using 8-nodes linear brick element. The model has 30 448 nodes, 24 693 elements, and 720 contact elements (only for the target contact interface). The mean area of contact elements is about 0.50 mm$^2$. The interaction between contact surfaces was simulated using the 'Lagrange multipliers' friction formulation with a friction coefficient of $\mu = 0.36$ (extracted from experimental results). Here, it is assumed that the friction coefficients of all contact surfaces are the same. In the FE model, the plastic deformation of the specimens is not considered, and its material parameters are as follows: 200 GPa modulus of elasticity and 0.3 Poisson's ratio. A quasi-static contact analysis was conducted to simulate the force transfer at the joint, where the moving specimen was displaced with a cyclic motion.

Figure 2.15 illustrates the comparison between the initial slope of the force–displacement curve obtained from the FE simulation and the experimental counterpart. The results show that both agree well in the stuck and gross slip regimes, which further verifies the effectiveness of the experimental results. In addition, the

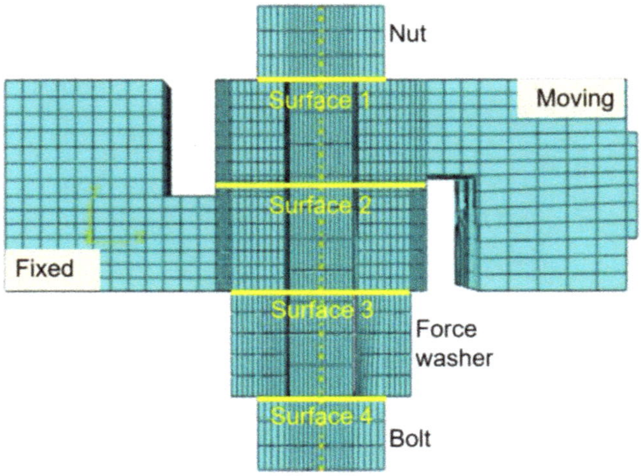


**Figure 2.14.** Finite element model of the bolt connection (the solid yellow lines indicate contact surfaces). (Adapted with permission from [1]. Copyright 2020 Elsevier.)

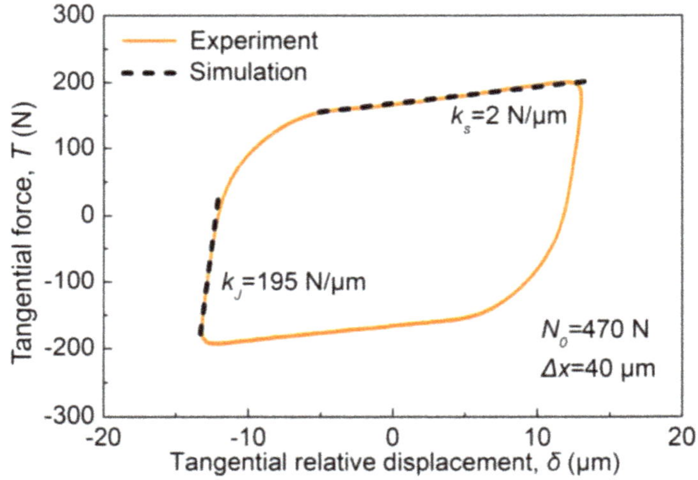


**Figure 2.15.** Comparison between the simulated stiffness and the experimental counterpart. (Reproduced with permission from [1]. Copyright 2020 Elsevier.)

simulation results also indicated the phenomenon of residual stiffness (a non-Coulomb friction behavior). To explain this phenomenon, a simplified 5 degrees of freedom lumped-parameter model is given in the next section.

### 2.3.2 Lumped-parameter model of bolted joint

Figure 2.16(a) depicts the simplified 5 degrees of freedom model of the bolt connection, where all components are assumed to be rigid and the contact interaction of the four contact surfaces—between the nut and the moving specimen,

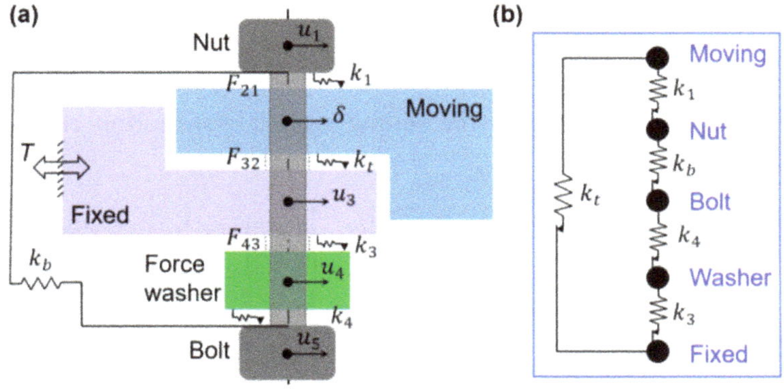


**Figure 2.16.** (a) A simplified 5 degree of freedom model of the bolt joint where the contact is simulated by the stick/slip contact element and (b) the rearrangement of the model for clarity. (Adapted with permission from [1]. Copyright 2020 Elsevier.)

the washer and the fixed specimen, the bolt head and the washer in addition to the target contact surface—are simulated using the ideal Coulomb contact element. The contact element consists of a linear spring and a slider in series, which can reproduce the stick and gross slip behaviors but not partial slip behavior at contact interfaces. This simplified model has five degrees of freedom in the horizontal direction of motion: $u_1$, $\delta$, $u_3$, $u_4$, and $u_5$.

The motion equations of the model when the target interface is in the stick and sliding regimes are derived. According to the FE analysis, regardless of whether the target interface is in the stick or sliding regime, all other contact interfaces are in a stick regime. Of course, this only applies to the load conditions in this example. When the target interface is in the stick regime, the motion equations of the model are expressed as

$$\begin{bmatrix} k_b + k_1 & -k_1 & 0 & 0 & -k_b \\ -k_1 & k_1 + k_t & -k_t & 0 & 0 \\ 0 & -k_t & k_t + k_3 & -k_3 & 0 \\ 0 & 0 & -k_3 & k_3 + k_4 & -k_4 \\ -k_b & 0 & 0 & -k_4 & k_b + k_4 \end{bmatrix} \begin{Bmatrix} u_1 \\ \delta \\ u_3 \\ u_4 \\ u_5 \end{Bmatrix} = \begin{Bmatrix} 0 \\ T \\ -R \\ 0 \\ 0 \end{Bmatrix}, \tag{2.7}$$

where $k_1$, $k_t$, $k_3$, and $k_4$ are the contact stiffness at the nut, at the target surface, at the washer and at the bolt head respectively, $k_b$ the stiffness between the bolt head and the nut, $R$ the reaction on the support. The solution of equation (2.7) is

$$\begin{Bmatrix} u_1 \\ T \\ R \\ u_4 \\ u_5 \end{Bmatrix} = \frac{\delta}{k} \begin{Bmatrix} k_1(k_3k_4 + k_3k_b + k_4k_b) \\ k_1k_tk_3k_4 + k_1k_tk_3k_b + k_1k_tk_4k_b + k_1k_3k_4k_b + k_tk_3k_4k_b \\ k_1k_tk_3k_4 + k_1k_tk_3k_b + k_1k_tk_4k_b + k_1k_3k_4k_b + k_tk_3k_4k_b \\ k_1k_4k_b \\ k_1k_b(k_3 + k_4) \end{Bmatrix}, \tag{2.8}$$

where

$$k = k_1k_3k_4 + k_1k_3k_b + k_1k_4k_b + k_3k_4k_b. \tag{2.9}$$

When the target interface is in the sliding regime, the motion equations of the model are expressed as

$$\begin{bmatrix} k_b + k_1 & -k_1 & 0 & 0 & -k_b \\ -k_1 & k_1 & 0 & 0 & 0 \\ 0 & 0 & +k_3 & -k_3 & 0 \\ 0 & 0 & -k_3 & k_3 + k_4 & -k_4 \\ -k_b & 0 & 0 & -k_4 & k_b + k_4 \end{bmatrix} \begin{Bmatrix} u_1 \\ \delta \\ u_3 \\ u_4 \\ u_5 \end{Bmatrix} = \begin{Bmatrix} 0 \\ T - \mu N_0 \\ \mu N - R \\ 0 \\ 0 \end{Bmatrix} \tag{2.10}$$

and the corresponding solution is

$$\begin{Bmatrix} u_1 \\ T \\ R \\ u_4 \\ u_5 \end{Bmatrix} = \frac{1}{k} \begin{Bmatrix} k_1\delta(k_3k_4 + k_3k_b + k_4k_b) \\ \mu N_0k + k_1k_3k_4k_b\delta \\ \mu N_0k + k_1k_3k_4k_b\delta \\ k_1k_4k_b\delta \\ k_1k_b(k_3 + k_4)\delta \end{Bmatrix}. \tag{2.11}$$

The comparison in figure 2.15 has demonstrated the reliability of the FE model of the joint in calculating the contact stiffness $k_J$, so the same FE model is utilized to determine the contact stiffness $k_1$, $k_t$, $k_3$, and $k_4$. The bending stiffness $k_b$ of the bolt shank is obtained using an analytical model [12]. This model considers the bolt shank as a beam. One end of the beam, the screw head, is clamped. The other end, the nut, restricted from rotation, but can freely move in the transverse direction. The stiffness is determined by the ratio of the transverse force in the clamp to the displacement of the nut,

$$k_b = \frac{12E_sI_z}{(1 + \Phi)L_s^3}, \tag{2.12}$$

where $L_s$, $E_s$, and $I_z$ are the bolt shank length, the elastic modulus of the bolt shank, and the second moment of area of the bolt shank, respectively. $\Phi$ is expressed as

$$\Phi = \frac{12E_sI_z}{GA_sL_s^2}, \tag{2.13}$$

where $G$ denotes the shear modulus of the material of the bolt, and $A_s$ the cross-section area of the bolt shank.

Table 2.2 lists the contact stiffness calculated from the finite element model and the bending stiffness of the bolt shank calculated from the analytical model. Figure 2.16(b) shows the equivalent model of the lumped-parameter model, which depicts the stiffness between the fixed and moving specimens, that is, the stiffness $k_J$ between the displacement measurement points. It can be seen from figure 2.16(b)

**Table 2.2.** Assessment of the repeatability of contact parameters and energy dissipation per cycle. (Reproduced with permission from [1]. Copyright 2020 Elsevier.)

| Stiffness | $k_1$ | $k_3$ | $k_4$ | $k_b$ | $k_J$ |
|---|---|---|---|---|---|
| Value (N $\mu$m$^{-1}$) | 16.0 | 15.2 | 16.0 | 5.0 | 195.0 |

that the overall joint stiffness $k_J$ equals the total stiffness of the target interface stiffness $k_t$ in parallel with other series stiffnesses $k_s$,

$$k_J = k_t + k_s \tag{2.14}$$

in which $k_s$ is expressed as

$$k_s = \frac{k_1k_3k_4k_b}{k_1k_3(k_4 + k_b) + k_4k_b(k_1 + k_3)}. \tag{2.15}$$

Substituting the stiffness value listed in table 2.2 into equation (2.14) obtains $k_s = 2.5$ N $\mu$m$^{-1}$, which corresponds to the residual stiffness in the experimental results. The contact stiffness of the target interface can be calculated through equation (2.15), $k_t = 192.5$ N $\mu$m$^{-1}$, which is in good agreement with the finite element results. Therefore, it is proved that the contact stiffness of the target interface can be extracted through the above method.

### 2.3.3 Correction of experimental results

According to the above stiffness analysis, a correction parameter $\Delta F$ is defined to correct the measured tangential friction force $T$, and the corrected result is the friction force $F_{32}$ of the target interface. This correction parameter is defined as the difference between the two. According to the aforementioned equations (2.7)–(2.11), it is written as

$$\Delta F = T - F_{32} = \frac{k_1k_3k_4k_b\delta}{k_1k_3k_4 + k_1k_3k_b + k_1k_4k_b + k_3k_4k_b}. \tag{2.16}$$

It can be seen from equation (2.16) that the correction parameter is linearly related to the relative displacement between the target surfaces, and has nothing to do with the motion state of the target interface. Figure 2.17 shows the comparison between the experimental and the corrected hysteresis loops with the corrected tangential friction held constant in the gross slip regime.

In order to reveal the sensitivity of the correction parameter to bolt preload and bolt shank length, a parametric analysis was performed. The stiffness values for each set of parameters were recalculated through the above-mentioned FE model and the beam analytical formula. Figure 2.18 plots the friction correction value as a function of tangential relative displacement under different bolt shank lengths and bolt preloads, which shows that the higher the compliance of the bolt shank, the smaller the correction of the measured contact force.

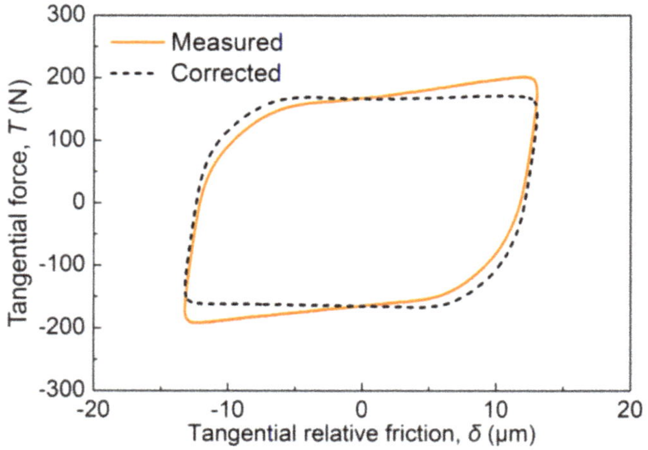


**Figure 2.17.** Comparison between the corrected and the measured hysteresis loops. (Reproduced with permission from [1]. Copyright 2020 Elsevier.)

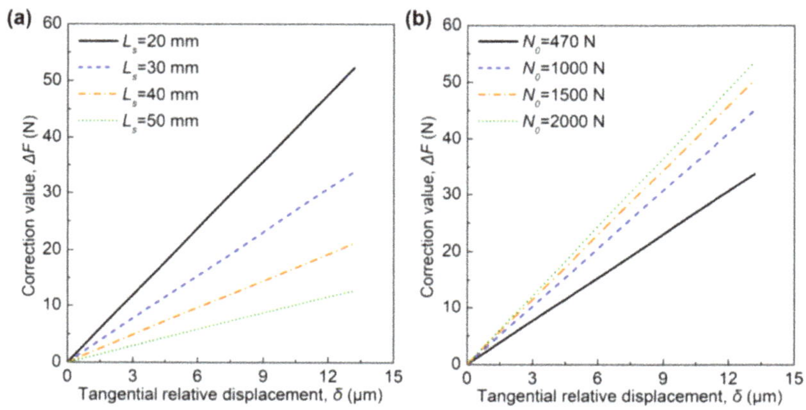


**Figure 2.18.** Friction correction value as a function of tangential relative displacement (a) under different bolt shank lengths and (b) under different bolt preloads. (Reproduced with permission from [1]. Copyright 2020 Elsevier.)

## 2.4 Summary

This chapter introduces a novel fretting apparatus to study friction hysteresis behavior in bolted joints. Compared with those test rigs in the literature, this device has some unique innovations.

1. This device uses only one laser vibrometer and a prism to measure the relative displacement of the contact surface, which is not only simple and easy to implement, but also cost-effective.
2. From the mechanical point of view, the device is carefully designed to allow accurate measurement of the interface hysteresis curve. It uses a symmetric support frame, making the device compact and self-balancing; in addition, the

leaf spring can well avoid out-of-phase vibration and guide the movement of the contact surface. All these detailed designs enable this test rig to meet high precision requirements and make it unique among test devices to date.

Experimental results demonstrate good repeatability of the test rig and show that it is sensitive to the excitation frequency. To further improve the confidence in the measured data, an additional investigation of the accuracy of the measurement was performed. A FE model and a simplified model of the bolt joint were given to analyze the load transfer, which revealed the origin of the residual stiffness in the gross slip regime. This chapter introduces some aspects that need to be paid attention to in terms of device design, experimental data analysis, and measurement accuracy, when testing the friction contact behavior of bolted joint interfaces, which can be used as an important reference.

## Bibliography

[1] Li D, Xu C, Botto D, Zhang Z and Gola M 2020 A fretting test apparatus for measuring friction hysteresis of bolted joints *Tribol. Int.* **151** 106431

[2] Eriten M, Polycarpou A A and Bergman L A 2011 Development of a lap joint fretting apparatus *Exp. Mech.* **51** 1405–19

[3] Brake M R W 2017 A reduced Iwan model that includes pinning for bolted joint mechanics *Nonlinear Dyn.* **87** 1335–49

[4] Abad J, Medel F J and Franco J M 2014 Determination of Valanis model parameters in a bolted lap joint: Experimental and numerical analyses of frictional dissipation *Int. J. Mech. Sci.* **89** 289–98

[5] Li Y, Hao Z, Feng J and Zhang D 2017 Investigation into discretization methods of the six-parameter Iwan model *Mech. Syst. Sig. Process.* **85** 98–110

[6] Abad J, Franco J M, Celorrio R and Lezáun L 2012 Design of experiments and energy dissipation analysis for a contact mechanics 3D model of frictional bolted lap joints *Adv. Eng. Software* **45** 42–53

[7] Schwingshackl C W, Petrov E P and Ewins D J 2012 Measured and estimated friction interface parameters in a nonlinear dynamic analysis *Mech. Syst. Sig. Process.* **28** 574–84

[8] Botto D and Lavella M 2014 High temperature tribological study of cobalt-based coatings reinforced with different percentages of alumina *Wear* **318** 89–97

[9] Botto D, Umer M, Gastaldi C and Gola M M 2017 An experimental investigation of the dynamic of a blade with two under-platform dampers *ASME Turbo Expo 2017: Turbomachinery Technical Conf. and Expo. (Charlotte, NC, 26–30 June)* (New York: The American Society of Mechanical Engineers)

[10] Umer M and Botto D 2019 Measurement of contact parameters on under-platform dampers coupled with blade dynamics *Int. J. Mech. Sci.* **159** 450–8

[11] Botto D and Umer M 2018 A novel test rig to investigate under-platform damper dynamics *Mech. Syst. Sig. Process.* **100** 344–59

[12] Przemieniecki J S 1968 *Theory of Matrix Structural Analysis* (New York: McGraw-Hill)

**IOP** Publishing

Fretting Friction and Wear at Bolted Joint Interfaces

**Chao Xu and Dongwu Li**

# Chapter 3

# Fretting wear of bolted joint interfaces under harmonic vibrations

Under transverse harmonic vibration, the long-term accumulation of interface friction behavior will inevitably lead to the wear of bolt connection surfaces and changes in the interface mechanical behavior. In this chapter, the effects of bolt preload and tangential excitation amplitude on the interface fretting behavior are first presented. Then, the evolution of the hysteresis behavior, contact parameters (including tangential contact stiffness and friction coefficient), and bolt preload, during a 12 h wear test period is investigated. The main physical mechanisms leading to this evolution are also analyzed.

## 3.1 Fretting wear experimental method

This section uses the fretting test apparatus described in chapter 2 to study the wear behavior of bolted joint interfaces under transversal harmonic vibrations [1]. In order to ensure stable excitation amplitude during the wear test, the piezoelectric actuator was closed-loop controlled using a built-in SGS sensor and servo control system. The friction force and tangential relative displacement were measured to characterize the interface friction hysteresis. The tangential contact stiffness and friction coefficient were extracted from the force–displacement curves. Given that interface wear is one of the important reasons why the bolt preload is loosened, the evolution of the bolt preload with increasing wear was also recorded.

### 3.1.1 Bolted joint specimens

The bolted joint specimens were made of ASTM 304 stainless steel. The shape and dimensions of the specimens are shown in figure 3.1. All joint specimens were manufactured using wire cutting technology, which results in a large surface roughness. To compare the influence of surface roughness on the results, all specimens were divided into two groups. The contact surfaces of the first group

doi:10.1088/978-0-7503-6214-6ch3  

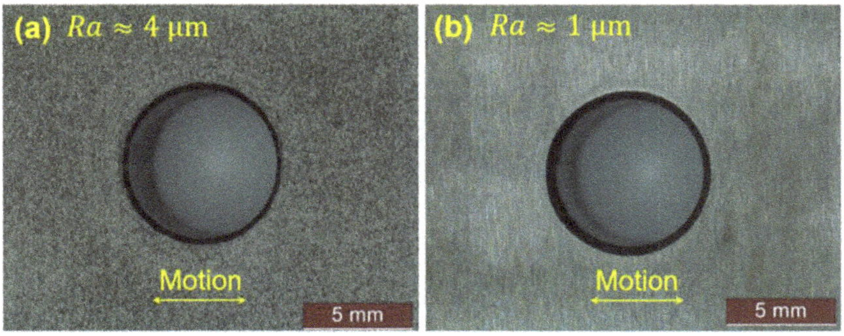


**Figure 3.1.** Photograph of the contact surfaces and corresponding surface roughness $R_a$, (a) rough surface: $R_a \approx 4\ \mu\text{m}$ and (b) smooth surface: $R_a \approx 1\ \mu\text{m}$ . (Reproduced with permission from [1]. Copyright 2020 Elsevier.)

**Table 3.1.** Summary of the wear test plan. (Reproduced with permission from [1]. Copyright 2020 Elsevier.)

| | Test #1/Test #2 | Test #3/Test #4 |
|---|---|---|
| Material | Stainless steel | Stainless steel |
| Roughness, $R_a$ | 4 $\mu$m/1 $\mu$m | 4 $\mu$m/1 $\mu$m |
| Excitation amplitude, $\Delta x$ | 50 $\mu$m | 40 $\mu$m |
| Excitation frequency, $f$ | 25 Hz | 25 Hz |
| Bolt preload, $N_0$ | 720 N | 720 N |
| Running time | 12 h | 12 h |
| Temperature | 25 ⍰ | 25 ⍰ |

specimens were carefully hand-polished using two different grades of sandpaper (first 800 grit and then 1200 grit), resulting in a roughness $R_a$ of about 1 $\mu$m. The second group of specimens were left without any treatment, and the roughness of the contact surfaces was approximately 4 $\mu$m.

### 3.1.2 Wear test plan

Table 3.1 summarizes the wear test specifications and operating conditions for four-group tests. The four groups of joint specimens were tested to study the effects of transversal excitation amplitudes and surface roughness on the interface fretting wear behavior. The average roughness of the contact surfaces of each joint specimen is listed in table 3.2. Given that too an high excitation amplitude (the maximum output displacement amplitude of the piezoelectric actuator is 70 $\mu$m) will cause the piezoelectric actuator to be damaged due to an excessive temperature due to long-lasting work, two nominal excitation amplitudes, $\Delta x = 50$ and 40 $\mu$m, were applied to the contact interfaces.

**Table 3.2.** Contact surface roughness of bolted joint specimens. (Reproduced with permission from [1]. Copyright 2020 Elsevier.)

| | Test #1 | Test #2 | Test #3 | Test #4 |
|---|---|---|---|---|
| Surface roughness of fixed specimen | 4.34 | 0.78 | 4.27 | 0.81 |
| Surface roughness of moving specimen | 5.19 | 0.92 | 4.43 | 0.90 |

In the analysis of experimental results, it is necessary to identify the friction coefficient of contact interfaces. The identification of friction coefficient depends on the friction force during the gross slip phase. Therefore, to directly extract the friction coefficient from the measured hysteresis curves, the bolt pretension force was set to 720 N such that the excitation amplitude enables the gross slip of joint interfaces. Different from the torque control method, the bolt preload can be directly measured using a force washer with great accuracy. Despite this, the process of tightening the bolts may lead to a 5% scattering in the bolt preload among different tests.

All wear tests last for 12 h (generating 1.08 million wear cycles). Due to the large amount of test data and limited hard disk memory, it is impossible to record them completely, so the following data acquisition strategy was used: for the first 20 min, 1 s data every 5 s was recorded; from 20 to 90 min, 1 s data every 40 s was recorded; from 90 to 720 min, 1 s data every 200 s was recorded. Data acquisition was implemented via LabVIEW code. The sampling frequency was set to 5 kHz, and no filtering was used.

Before and after each test, the bolted joint specimens were placed in an ultrasonic cleaning machine and cleaned with alcohol for 30 min. This process can minimize the impact of particles and engine oil on test results. Subsequently, they were dried in a temperature-controlled box to guarantee a dry contact condition. After cleaning and drying, the contact surfaces were photographed using a microscope (Leica S9D stereomicroscope). During the test, the ambient temperature was 25 °C and the humidity was maintained at ~30%. The detailed experimental procedure can be found in [1].

## 3.2 Friction hysteresis curves

Before the long-term wear test, two sets of experiments on bolted joint specimens with smooth surfaces ($R_a \approx 1\ \mu$m) to study the influence of bolt preload and excitation amplitude on the interface friction hysteresis behavior and contact parameters.

### 3.2.1 Effect of bolt preload

Figure 3.2 plots the tangential force as a function of tangential relative displacement under different bolt preloads with the same sinusoidal excitation (25 $\mu$m amplitude and 25 Hz frequency). Although the applied excitation amplitude is the same among different tests, the measured displacement amplitudes at the contact interface differ

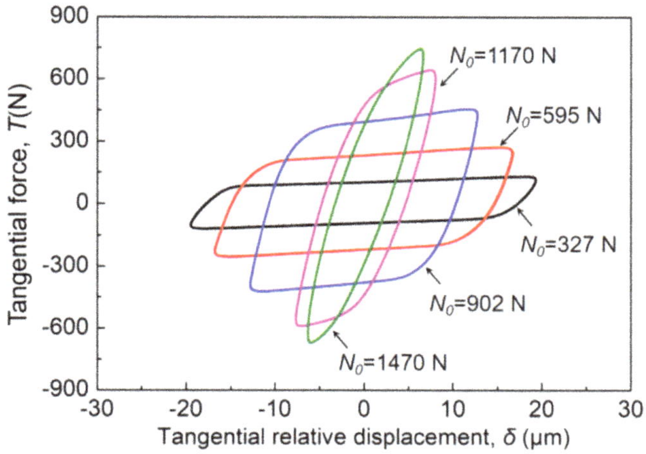


**Figure 3.2.** Measured friction hysteresis loops under different bolt preloads and the same nominal displacement imposed by the actuator. (Adapted with permission from [5]. Copyright 2020 Elsevier.)

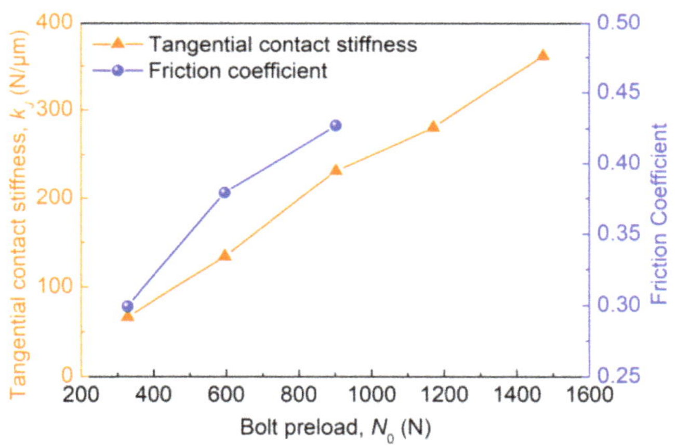


**Figure 3.3.** Tangential contact stiffness and friction coefficient of joint interfaces for different bolt preloads. (Adapted with permission from [5]. Copyright 2020 Elsevier.)

due to the semi-closed-loop control of the test rig. The results show a significant impact of bolt preload on the shape of the hysteresis loops. As the bolt preload increases, the relative displacement amplitude gradually decreases, while the friction force amplitude increases, and the shape of the hysteresis loop changes from a parallelogram-like to an elliptical-like.

Figure 3.3 shows the tangential contact stiffness and friction coefficient of joint interfaces for different bolt preloads, which were extracted using the method given in section 2.2.1. It can be seen that the tangential contact stiffness ramped up with the increase of bolt preload. From the perspective of rough contact, the surface of mechanical parts is composed of a large number of asperities of varying sizes. According to the rough contact theory [2], a larger bolt preload will result in a larger actual contact area on the interface. On the other hand, Mindlin's contact theory [3] also indicates that the tangential contact stiffness depends proportionally on the contact area. Therefore, the evolution of tangential contact stiffness is expected.

The friction coefficient and tangential contact stiffness show a similar evolutionary trend with the increase of bolt preload. The reason why the friction coefficients at 1170 and 1470 N preloads were not given is because the interface gross slip was not reached in both cases. Generally speaking, without considering wear, the friction coefficient does not change with external loads for materials with the same contact surface conditions. One explanation for the 'unusual' change in friction coefficient is that the process of applying bolt preload changed the morphology of the contact surface. Similar experimental results also can be found in [4].

### 3.2.2 Effect of excitation amplitude

Joint specimens were tested with constant bolt preload and different displacement amplitudes to investigate the effect of excitation amplitudes. After each test, the device was disassembled and reassembled to release the interface force and to ensure that the testing conditions remain unchanged between different tests. Figure 3.4 depicts hysteresis loops under different nominal displacement amplitudes and the same bolt preload. When the imposed displacement amplitudes increased, the hysteresis loop changed from an ellipse to a parallelogram, accompanied by the emergence of gross slip at the interface.

In addition, at the end of the gross slip phase during reloading, a sudden uplift in friction force can be observed, as shown in the solid purple ellipse of figure 3.4(b). However, this phenomenon did not appear during the unloading stage, and in the hysteresis loops at small displacement amplitudes. Therefore, it is speculated that this phenomenon is caused by the unexpected contact between the bolt shank and the bolt hole. This unexpected contact is due to the misalignment (non-coaxial) of

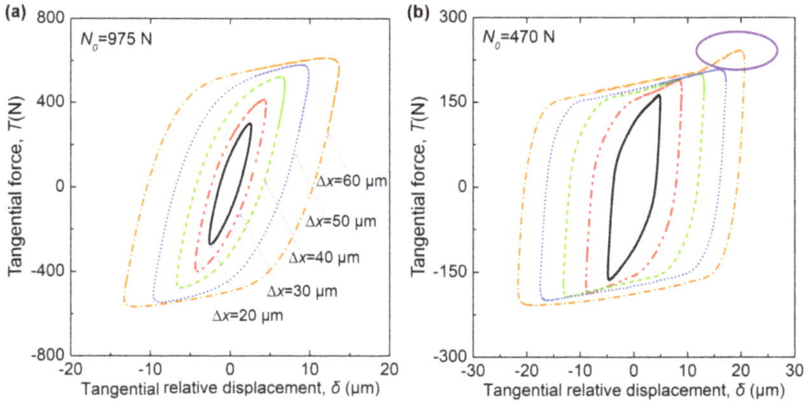


**Figure 3.4.** Measured friction hysteresis loops under different nominal displacement amplitudes imposed by the actuator and the same bolt preload: (a) $N_0 = 975$ N and (b) $N_0 = 470$ N. (Adapted with permission from [5]. Copyright 2020 Elsevier.)

the bolt shank with the center of the bolt holes. It should be noted that the 'misalignment' here is not the misalignment of the bolt holes and has been avoided well using a feeler gauge in the tests. Figure 3.5 depicts the alignment case (perfect assembly) and misalignment case (non-coaxial) of the bolt shank and the bolt hole, respectively. In the alignment case (figure 3.5(a)), the axis of the bolt shank coincides with the axis of the bolt hole of the specimen, so that as long as the relative displacement between the interfaces is lower than the gap between the bolt shank and the hole wall, there will be no contact between the two. In the misalignment case (figure 3.5(b)), the axis of the bolt shank is offset from the bolt hole axis due to installation error. Therefore, if the imposed displacement amplitude is greater than the minimum gap, the bolt shank will come into contact with the bolt hole. Moreover, when the displacement amplitude is lower than the maximum gap along the motion direction, the bolt shank and the hole wall will no longer contact during the reverse movement stage. Since it is difficult to control the relative position of the shank and the bolt hole during tightening of the bolt, no repeated experiments were conducted here to verify the above inference.

Figure 3.6 plots the tangential contact stiffness as a function of nominal imposed displacement amplitude. The results show that the tangential contact stiffness was not evidently sensitive to the excitation amplitude.

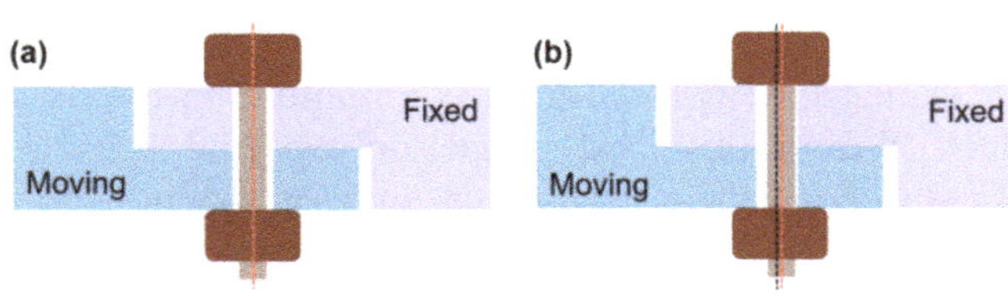


**Figure 3.5.** The assembly position of the bolt shank and the bolt hole (a) alignment (perfect assembly) and (b) misalignment (non-coaxial).

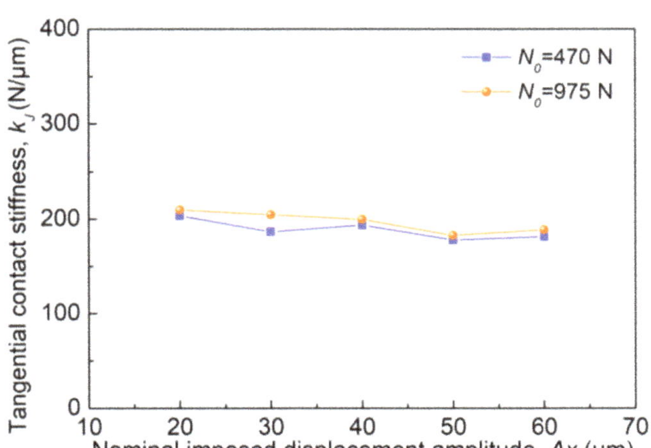


**Figure 3.6.** Tangential contact stiffness as a function of nominal imposed displacement amplitude under different bolt preloads. (Adapted with permission from [5]. Copyright 2020 Elsevier.)

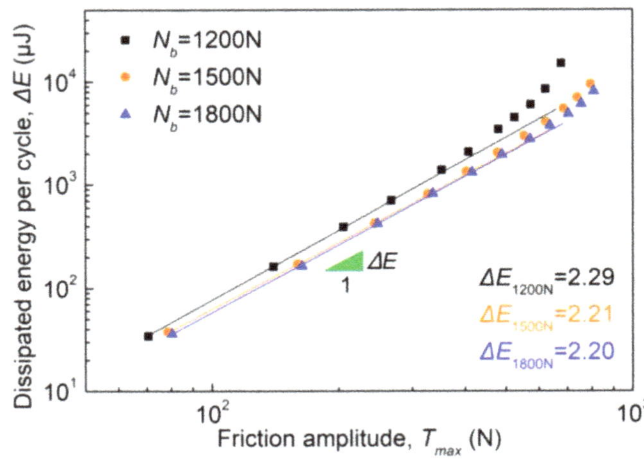


**Figure 3.7.** Friction-induced energy dissipation per cycle as a function of friction amplitude. (Adapted with permission from [5]. Copyright 2020 Elsevier.)

Figure 3.7 shows the relation between the dissipated energy per cycle induced by interface friction and the friction force amplitudes under three different bolt preloads (1200, 1500, and 1800 N). All data are plotted in a double logarithmic coordinate system. When the friction amplitude is below about 500 N, the dissipated energy per cycle indicates a scaling law with the friction amplitude in the double logarithmic coordinate system. Within this load range, only partial slip can occur at the interface. The fitted slope of the energy–force curves is around 2.25 and falls within the range of 2–3 as found in the literature [6]. Furthermore, the energy–force power-law relationship is not sensitive to the bolt preload.

## 3.3 Wear test results and discussion

This section will show the evolution of hysteresis curves, bolt preload, contact parameters, and surface morphology during long-term wear tests. The mechanism underlying these evolutionary laws will also be revealed.

### 3.3.1 Evolution of hysteresis loops

Figure 3.8 shows the evolution of hysteresis loops measured during wear tests. It can observed that as the wear cycles increased, the shape of the hysteresis loop changed. Also, relative to the results of smooth contact surfaces (tests #2 and #4), the variation in the hysteresis loops was more significant for rough surfaces (tests #1 and #3). There are two main reasons for this evolution. The first reason is the change in contact surface morphology caused by wear, which in turn causes the variation of tangential contact stiffness and friction coefficient. The second one is the change in bolt preload due to long-term transversal vibration.

In tests #1 and #3, as the wear cycle increased, the tangential force during the gross slip stage dropped significantly and the sliding stroke at the interface gradually increased. The 'warping' phenomenon (stiffness hardening) occurred at the end of

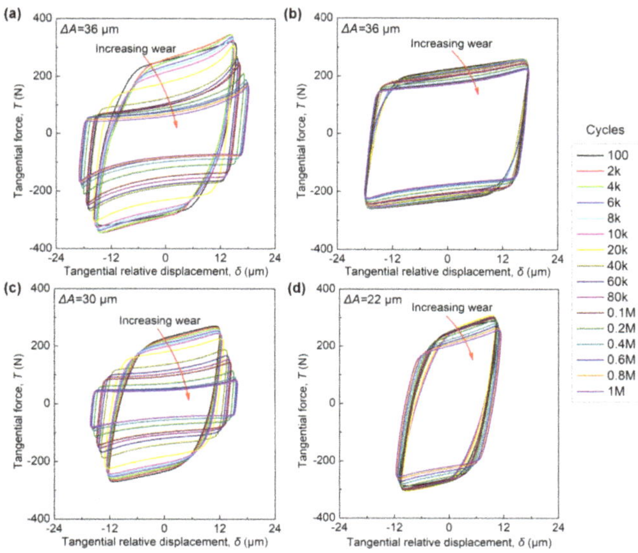


**Figure 3.8.** Evolution of hysteresis loops with increasing wear in four tests, (a) test #1: $\Delta x = 50$ $\mu$m, $R_a \approx 4$ $\mu$m, (b) test #2: $\Delta x = 50$ $\mu$m, $R_a \approx 1$ μm, (c) $\Delta x = 40$ $\mu$m, $R_a \approx 4$ $\mu$m, and (d) $\Delta x = 40$ $\mu$m, $R_a \approx 1$ $\mu$m. $\Delta A$ represents the average sliding strokes at the interface. (Adapted with permission from [1]. Copyright 2020 Elsevier.)

the gross slip phase of the hysteresis curve. For test #2, the sliding friction force decreased with the wear cycles, but the decline was slight. The partial slip stage in the hysteresis loop evolved from visible to almost disappearing. Moreover, after about 0.1 million wear cycles, the 'warping' phenomenon also occurred, but the sliding strokes did not change evidently. The results for test #4 show a similar evolutionary trend as test #2, nevertheless, the difference is that there was no obvious 'warping' at the end of the gross slip stage and a slight increase in the relative displacement amplitude.

In tests #1 and #2, the average sliding stroke $\Delta A$ (i.e. twice the tangential relative displacement amplitude) was 36 $\mu$m; whereas in tests #3 and #4, this value was 30 $\mu$m and 22 $\mu$m, respectively. In addition, the sliding stroke on rough surfaces (tests #1 and #3) was more dispersed than on smooth surfaces (tests #2 and #4).

Figure 3.9 plots the normalized hysteresis loops in which the tangential force was divided by the bolt preload. The overall evolutionary trend was opposite to the trend shown in figure 3.8 and the normalized tangential force gradually raised with increasing wear. The normalized tangential force is directly related to friction coefficient, so the evolution trend of the friction coefficient can be roughly seen from figure 3.9. It will be discussed in detail in section 3.3.3.

As shown in figure 3.9, the 'warping' phenomenon at the end of the gross slip regime was obvious, and the larger the relative displacement amplitude, the more evident this phenomenon was. The 'warping' phenomenon was also widely observed in the results of other fretting tests [7–9], but its underlying physical mechanism has not yet been fully revealed. There are two possible explanations: (i) interaction

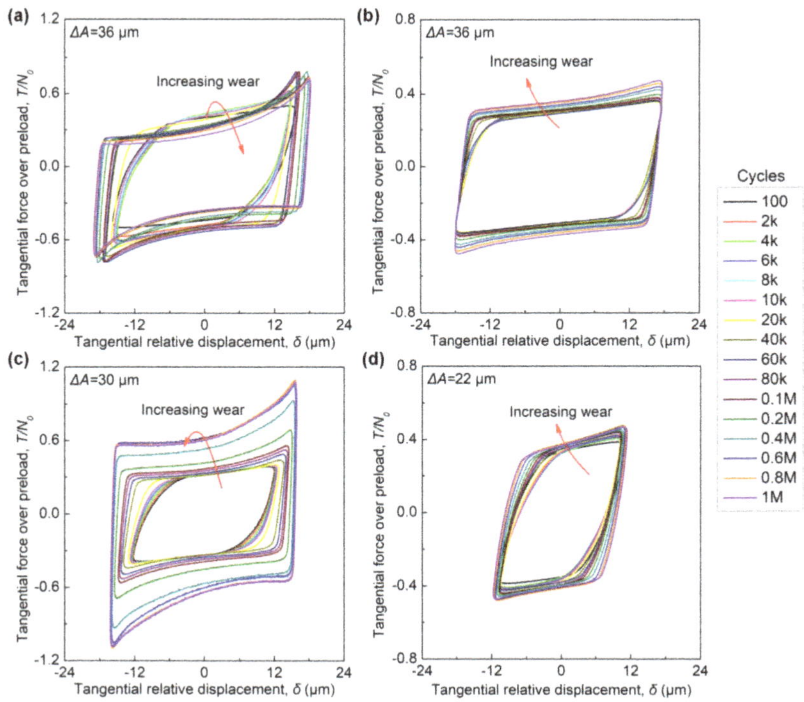


**Figure 3.9.** The ratio of tangential force to bolt preload as a function of tangential relative displacement for different tests: (a) test #1: $\Delta x = 50\ \mu m$, $R_a \approx 4\ \mu m$, (b) test #2: $\Delta x = 50\ \mu m$, $R_a \approx 1\ \mu m$, (c) $\Delta x = 40\ \mu m$, $R_a \approx 4\ \mu m$, and (d) $\Delta x = 40\ \mu m$, $R_a \approx 1\ \mu m$. (Adapted with permission from [1]. Copyright 2020 Elsevier.)

between wear scars on the contact surfaces and (ii) intermittent contact and separation between the bolt shank and the through hole.

Figure 3.10 plots the dissipated energy per cycle as a function of the cumulative dissipated energy. The cumulative dissipated energy refers to the sum of energy dissipation caused by interface friction before the current moment (that is, the sum of the areas of all hysteresis curves before the current moment), which is related to the time variable. It can be seen that the final cumulative dissipated energy on rough surfaces (tests #1 and #3) is significantly smaller than that on smooth surfaces (tests #2 and #4).

In tests #1 and #3, the dissipated energy per cycle first decreased noticeably and then gradually stabilized as the cumulative dissipated energy increased. The dissipated energy during the final wear cycle is only 42% and 25% of its initial value, respectively. The energy dissipation per cycle in tests #2 and #4 showed visible oscillations in the first thousands of wear cycles, and then gradually stabilized. The energy dissipation in the final state in test #2 was 84% of its initial value, while in test #4, except for the initial oscillation, it remained almost unchanged. The comparison shows that the dissipated energy per cycle of smooth contact surfaces (tests #2 and #4) was considerably larger than that of rough contact surfaces (tests #1 and #3), except for the first thousand wear cycles.

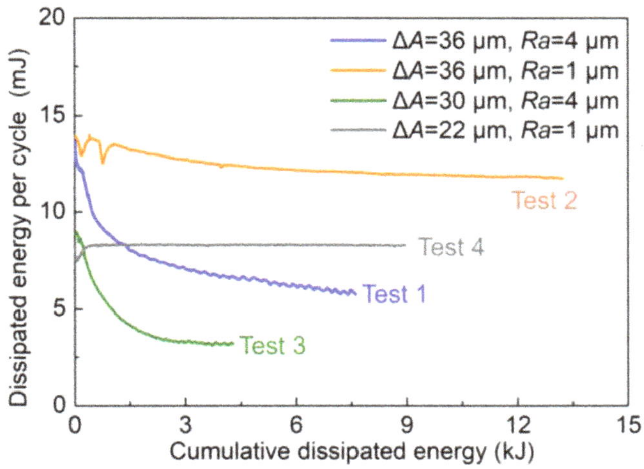


**Figure 3.10.** Dissipated energy per cycle as a function of cumulative dissipated energy for different tests. (Adapted with permission from [1]. Copyright 2020 Elsevier.)

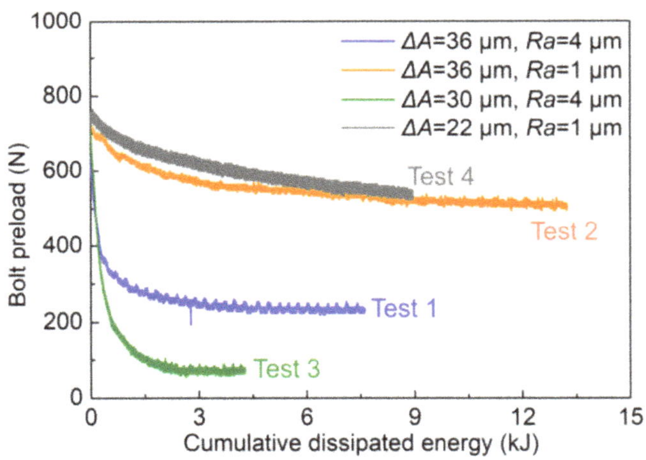


**Figure 3.11.** Variation of bolt preload with cumulative dissipated energy in different tests. (Adapted with permission from [1]. Copyright 2020 Elsevier.)

### 3.3.2 Evolution of bolt preloads

Fretting wear tests are usually conducted under a constant normal load, while the wear test in this chapter is different from those standard tests. Here the normal load exerted on the contact surface here was not constant but varied with increasing wear. Figure 3.11 plots the measured bolt preload as a function of the cumulative dissipated energy. In all tests, the bolt preload first experienced a sharp decline, and then gradually tended to an asymptotic steady state. Nonetheless, a clear difference can be observed that the bolt preload decreased more dramatically during the initial wear stage for rough surfaces compared to smooth surfaces.

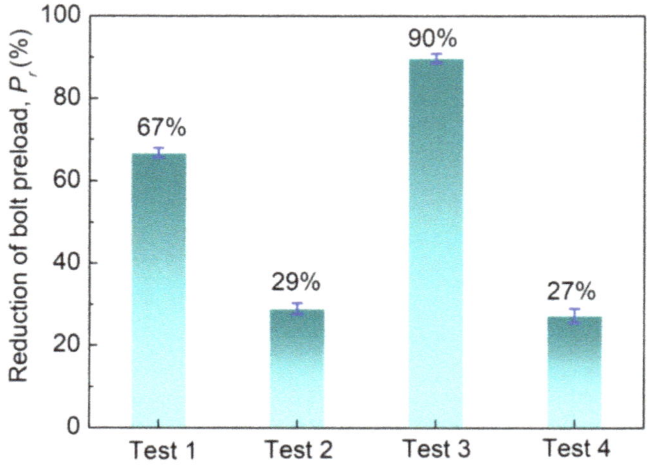


**Figure 3.12.** Reduction of bolt preload with respect to the initial value in different tests, $P_r = (N_{0\text{-initial}} - N_{0\text{-end}})/N_{0\text{-initial}}$, $N_{0\text{-initial}}$ is the initial value of the bolt preload, and $N_{0\text{-end}}$ its final value. The blue bar represents the standard deviation of $P_r$. (Adapted with permission from [1]. Copyright 2020 Elsevier.)

Figure 3.12 shows the relative reduction of bolt preload, $P_r$, in different tests. $P_r$ is defined as $(N_{0\text{-initial}} - N_{0\text{-end}})/N_{0\text{-initial}}$, where $N_{0\text{-initial}}$ is the initial value of the bolt preload, and $N_{0\text{-end}}$ its final value. The results show that the relative reduction of bolt preload for the rough contact surfaces (tests #1 and #3) was more evident than that for the smooth contact surfaces (tests #2 and #4). The decrease in bolt preload in test #3 even reached 90% of its initial value, while in both tests #2 and #4 this value was below 30%.

The attenuation of bolt preload under transversal vibration has been widely studied in the literature. A reasonable explanation for the loosening of the bolt preload is that during the fretting wear process, the transverse shearing action between contact pairs causes the peaks of surface asperities to be cut and ground into micro-particles. Then they are removed from the contact interface. This process reduces the interference fit between the contact interfaces, thereby causing the bolt preload to loosen. Experimental results also suggest that rough surfaces (with higher asperities prone to be cut) experienced a greater decline in bolt preload than smooth surfaces. Recent research claims that the main cause of bolt preload loosening at the early stage is stress release and stress redistribution of thread teeth [10]. However, the effect of joint surface roughness was not studied.

### 3.3.3 Evolution of contact parameters

It is well known that the fretting wear will seriously affect the tangential contact stiffness and friction coefficient of contact surfaces [11–13]. These two contact parameters can be identified from the measured hysteresis curve, referring to section 2.2.1.

Figure 3.13 plots the tangential contact stiffness as a function of the cumulative dissipated energy for different tests. It shows that the tangential contact stiffness

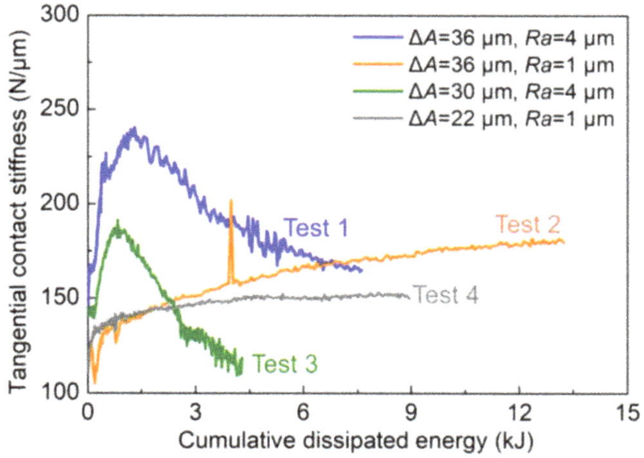


**Figure 3.13.** Tangential contact stiffness versus cumulative dissipated energy. (Adapted with permission from [1]. Copyright 2020 Elsevier.)

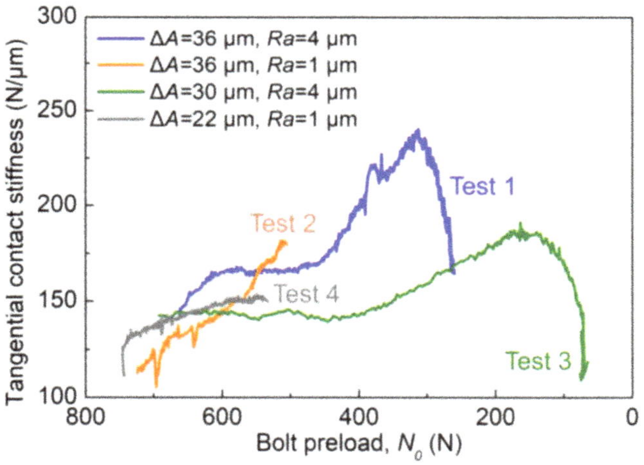


**Figure 3.14.** Tangential contact stiffness versus bolt preload. (Adapted with permission from [1]. Copyright 2020 Elsevier.)

underwent an evident variation during the whole test period. In tests #1 and #3, the tangential contact stiffness has a similar evolutionary trend: first rising steeply, and then reaching a peak at a cumulative dissipated energy of approximately 1.3 kJ, and finally declining slowly. Besides, throughout the wear test, the tangential contact stiffness of test #1 was greater than that of test #3. Figure 3.14 shows the relationship between tangential contact stiffness and bolt preload, indicating that although the bolt preload decreases with increasing wear, the tangential contact stiffness still presents an increasing trend.

Several examples of experimental evidence suggest that a greater normal load leads to a greater tangential contact stiffness. However, according to the classical contact theory—the Mindlin solution [14]—it can be concluded that the tangential contact stiffness is proportional to the radius of the contact area and is independent of the normal load, as following,

$$k_t = \frac{8Ga}{2 - \vartheta}, \tag{3.1}$$

where $G$ denotes the shear modulus of the material, $a$ the radius of contact area, and $\vartheta$ Poisson's ratio. Therefore, the relationship between normal load and tangential contact stiffness appears to be related to changes in contact area, i.e. increasing normal load causes the contact area to rise, thereby increasing tangential contact stiffness. In this chapter, the normal load was variable and diminished with the wear cycles, the tangential contact stiffness thus was expected to decrease. However, the interface wear results in an increase in the contact area and therefore increasing contact stiffness.

In addition, some studies have shown that the tangential contact stiffness is also related to the height of asperities on the contact surfaces [11, 15]. As can be seen in figure 3.13, the initial contact stiffness for the rough contact surfaces (tests #1 and #3) is much greater than the results for the smooth contact surfaces (tests #2 and #4). As a consequence, the increase in tangential contact stiffness may be mainly caused by the increased interaction between wear scars. This interaction increases the resistance to the relative motion between the contact surfaces during the stick phase. However, when the bolt preload drops to a certain level, the impact of the drop in bolt preload on the contact surface will exceed the impact of the interaction between the wear scars and become dominant, as shown in figure 3.14. This is why the contact stiffness decreases after reaching its peak value.

In summary, the evolution of tangential contact stiffness depends not only on the change of interface morphology caused by wear but also on the loosening of bolt preload. In different wear states, the dominant factors leading to changes in contact stiffness are different.

In tests #2 and #4, the tangential contact stiffness showed a rapid increase in early wear, and then gradually stabilized. The same trend can also be observed in other experiments [9, 11] where the normal load is constant during the wear tests and the roughness of the contact surfaces is about 1 $\mu$m. Therefore, the change in contact surface morphology is the main reason for the variation of the contact stiffness.

Figure 3.15 shows the friction coefficient as a function of the cumulative dissipated energy for different tests. In test #1, the friction coefficient first increased slightly and then decreased, accompanied by evident oscillations. In test #3, the friction coefficient peaked at a cumulative dissipated energy of about 3 kJ and then decreased slowly with fluctuations. The peak value of the friction coefficient even reached twice its initial value. The fluctuation in friction coefficient could be caused by wear particles. Specifically, the generation of abrasive particles leads to an augment in the friction coefficient, while the removal of abrasive particles results in its abatement.

Tests #2 and #4 show a completely different evolutionary trend in friction coefficients compared to tests #1 and # 3. The friction coefficient scaled up at the

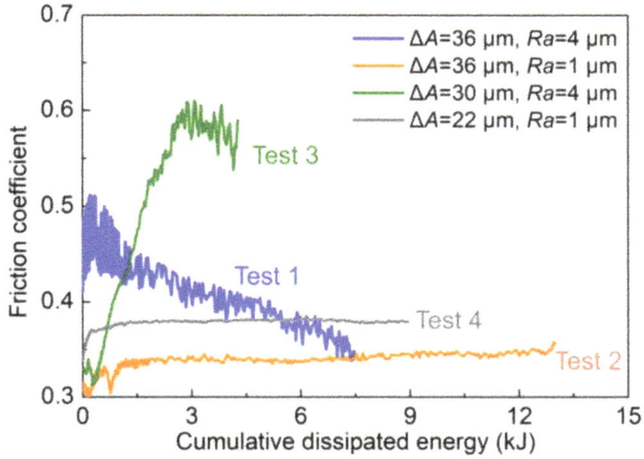


**Figure 3.15.** Variation of friction coefficient with cumulative dissipated energy in different tests. (Adapted with permission from [1]. Copyright 2020 Elsevier.)

initial wear stage and then stabilized. The same phenomenon also occurred in the tests in the literature [9, 16, 17]. As explained in [18], in the early stage of wear, the increment in friction coefficient is caused by the rapid generation of wear particles on the contact surfaces. As the wear deepens, the friction coefficient diminishes due to the weakening of the asperity deformation and the ploughing action. Finally, when the production and removal of wear particles reach a balance, the contact surfaces become smooth and the friction coefficient tends to a stable state.

Overall, the experimental results emphasize the significant impact of surface roughness on the evolution of friction coefficient. The surface with higher roughness shows a more evident variation of the friction coefficient than that with lower roughness. This is because higher roughness is more easily cut by shear loads and generates larger particles.

### 3.3.4 Worn surfaces

Figure 3.16 shows the microscopic images of contact surfaces after wear tests obtained using a steremicroscope in different tests. The area enclosed by the yellow curves is the wear scars in these images. For different tests, the distribution of wear scars is also different, with obvious uncertainties. In test #1, the wear scars are some vertical stripes evenly distributed along the direction of interface movement. Theses stripe-like scars coincides with the traces of wire cutting. Most likely, the manufacturing process left the joint surface with pronounced waviness. In test #3, the wear scars are mainly concentrated near the bolt holes, which may be due to local material accumulation and the protrusion of the hole edges caused by drilling. In tests #2 and #4, the wear scars are mainly distributed near the left boundary of the contact surface and only a small part near the right boundary. This is probably caused by hand polishing of contact surfaces.

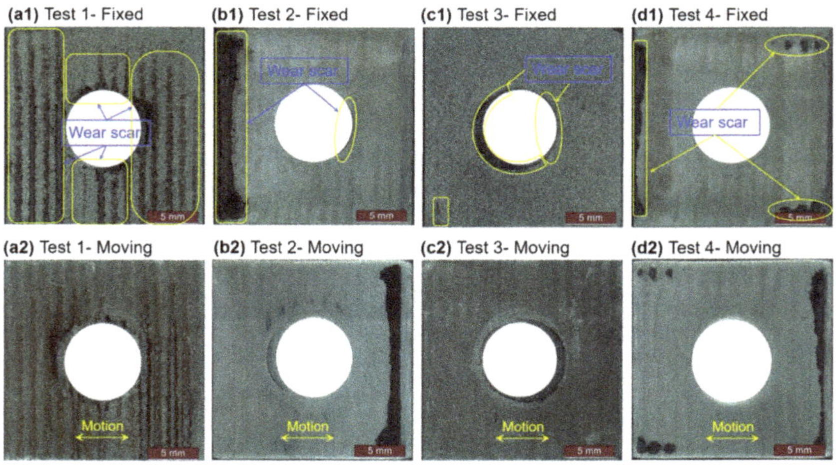


**Figure 3.16.** Contact surface images after wear tests: (a) test #1: $\Delta x = 50$ $\mu$m, $R_a \approx 4$ $\mu$m, (b) test #2: $\Delta x = 50$ $\mu$m, $R_a \approx 1$ $\mu$m, (c) $\Delta x = 40$ $\mu$m, $R_a \approx 4$ $\mu$m, and (d) $\Delta x = 40$ $\mu$m, $R_a \approx 1$ $\mu$m. (Adapted with permission from [1]. Copyright 2020 Elsevier.)

## 3.4 Summary

This chapter presents the measured force–displacement relationship and identified contact parameters for bolt joint interfaces in different fretting friction and wear scenarios, and discusses the causes and influencing mechanisms of hysteresis curve changes and contact parameter evolution. The main results and conclusions are as follows:

1. Under the condition of fretting friction, the shape of the hysteresis loop, reflecting the contact state of the joint interface, changes with the bolt preload and the tangential excitation amplitude. When the interface is in the partial slip state, there is a power-law relationship between the friction-induced energy dissipation per cycle and the friction amplitude, and the power exponent is about 2.2, which is smaller than that reported in the literature. If the bolt shank and the screw hole are not coaxial during the installation process, the two may come into unexpected contact, resulting in a sudden change in the tangential contact stiffness of the joint interface and an asymmetric hysteresis loop.
2. The fretting wear test results show that the interface hysteresis behavior and contact parameters change with increasing wear, accompanied by the attenuation of bolt preload. In particular, surface roughness has a significant influence on the wear behavior. Under the same sliding stroke, the change of hysteresis curve and contact parameter for the rough surface is more dramatic than that for the smooth surface. In general, the sliding force gradually decreases with the increase of interface wear.
3. It is well known that interface wear is one of the main reasons for bolt loosening, and the experimental results in this chapter also confirm this. Compared with smooth surfaces, the bolt preload on rough surfaces decays

faster and by a greater amount, even reaching 90% of the initial value. A reasonable explanation is given that the higher the height of the asperities on the surface, the easier they are to deform or be cut off under shearing action, thereby reducing the interference fit between the connection interfaces and causing a decrease in the bolt preload.

4. The evolution of tangential contact stiffness induced by interface wear is related to the change of surface morphology and bolt preload. These two factors are mutually opposed. Theoretically, when the interface morphology remains unchanged, the tangential contact stiffness is mainly related to the contact area, which in turn depends on the normal load. However, the experimental results show that as the bolt preload decreases, the contact stiffness increases, indicating that during the wear process, the increase in contact area caused by wear dominates the change of tangential contact stiffness. When the bolt preload drops to a very low value, the contact stiffness of rough surfaces peaks. However, since the preload of smooth surfaces does not drop to a very low level, it keeps increasing during the wear cycle.
5. There are significant differences in the evolution trends of the friction coefficient between the rough surface and the smooth surface. The friction coefficient of a smooth surface increases rapidly at first and then gradually stabilizes, while the friction coefficient of a rough surface increases at first and then decreases and gradually stabilizes. Rough surfaces are more likely to produce wear particles during the wear process. In the early stages of wear, the wear particles remaining on the contact surface are the main reason for the increase in the friction coefficient. When the generation and removal of wear particles on the contact surface reaches a certain balance, the friction coefficient gradually reaches a stable state.

## Bibliography


[1] Li D, Botto D, Xu C and Gola M 2020 Fretting wear of bolted joint interfaces *Wear* **458** 203411

[2] Bhushan B 1998 Contact mechanics of rough surfaces in tribology: multiple asperity contact *Tribol. Lett.* **4** 1–35

[3] Mindlin R D and Deresiewicz H 1953 Elastic spheres in contact under varying oblique forces *ASME J. Appl. Mech.* **20** 327–44

[4] Eriten M 2012 *Multiscale Physics-based Modeling of Friction* (Champaign, IL: University of Illinois at Urbana-Champaign)

[5] Li D, Xu C, Botto D, Zhang Z and Gola M 2020 A fretting test apparatus for measuring friction hysteresis of bolted joints *Tribol. Int.* **151** 106431

[6] Ames N M, Lauffer J P, Jew M D, Segalman D J, Gregory D L, Starr M J and Resor B R 2009 *Handbook on Dynamics of Jointed Structures* (Albuquerque, NM, and Livermore, CA: Sandia National Laboratories) No. SAND2009-4164

[7] Mulvihill D M, Kartal M E, Olver A V, Nowell D and Hills D A 2011 Investigation of non-Coulomb friction behaviour in reciprocating sliding *Wear* **271** 802–16

[8] Lavella M, Botto D and Gola M M 2013 Design of a high-precision, flat-on-flat fretting test apparatus with high temperature capability *Wear* **302** 1073–81

[9] Schwingshackl C W, Petrov E P and Ewins D J 2012 Measured and estimated friction interface parameters in a nonlinear dynamic analysis *Mech. Syst. Sig. Process.* **28** 574–84
[10] Gong H, Ding X, Liu J and Feng H 2022 Review of research on loosening of threaded fasteners *Friction* **10** 335–59
[11] Fantetti A, Tamatam L R, Volvert M, Lawal I, Liu L, Salles L, Brake M R W, Schwingshackl C W and Nowell D 2019 The impact of fretting wear on structural dynamics: experiment and simulation *Tribol. Int.* **138** 111–24
[12] Tamatam L R, Botto D and Zucca S 2021 A novel test rig to study the effect of fretting wear on the forced response dynamics with a friction contact *Nonlinear Dyn.* **105** 1405–26
[13] Li B, Li P, Zhou R, Feng X Q and Zhou K 2022 Contact mechanics in tribological and contact damage-related problems: a review *Tribol. Int.* **171** 107534
[14] Johnson K L 1987 *Contact Mechanics* (Cambridge: Cambridge University Press)
[15] Kartal M E, Mulvihill D M, Nowell D and Hills D A 2011 Measurements of pressure and area dependent tangential contact stiffness between rough surfaces using digital image correlation *Tribol. Int.* **44** 1188–98
[16] Hintikka J, Lehtovaara A and Mäntylä A 2015 Fretting-induced friction and wear in large flat-on-flat contact with quenched and tempered steel *Tribol. Int.* **92** 191–202
[17] Kartal M E, Mulvihill D M, Nowell D and Hills D A 2011 Determination of the frictional properties of titanium and nickel alloys using the digital image correlation method *Exp. Mech.* **51** 359–71
[18] Suh N P and Sin H C 1981 The genesis of friction *Wear* **69** 91–114

**Chao Xu and Dongwu Li**

# Chapter 4

# Fretting wear of bolted joint interfaces under random vibrations

In real engineering applications, external loads are usually random. This chapter introduces a test method of fretting wear at bolted joint interfaces under lateral random vibrations, the evolution law of wear-induced contact parameters, and the wear mechanism. The experimental results show that compared with the sinusoidal fretting test, the evolution law of the friction coefficient in the random fretting test is almost unchanged, but the contact stiffness is significantly different. Understanding these phenomena and the mechanisms behind them will help improve the design and durability of bolted connections in dynamic environments.

## 4.1 Fretting test method under random excitation

In random vibration both the amplitude and frequency are random, so the wear mechanism and wear rate involved will change compared to sinusoidal fretting tests. It is well known that a smaller sliding amplitude induces a partial slip regime in which the surface damage often tends to be a crack. In comparison, a larger sliding amplitude induces a gross slip regime, resulting in surface wear damage [1–3]. The randomness of the sliding amplitude complicates the fretting process due to the existence of a mixed-fretting regime and possible instability in the wear debris generation and ejection [4]. Numerous studies have shown that the variable sliding amplitude greatly affects the surface fretting stability and consequently the friction and wear mechanisms [5]. For example, for steel-to-steel contacts, a partial slip regime results in mild abrasive wear, while a gross slip regime results in abrasive wear, oxidative wear, and delamination [6].

To reproduce realistic fretting wear of bolted joint interfaces, a narrowband Gaussian process is used as input to the test apparatus introduced in chapter 2. Note that this chapter focuses on the evolution of contact parameters of bolted joint interfaces under transverse random vibrations, but does not attempt to analyze the

doi:10.1088/978-0-7503-6214-6ch4  

changes in wear volume and wear rate. Certainly, in the analysis of wear results, work rate is an important concept for evaluating wear. Work rate is usually used to relate the wear volume to the integral of the applied load and the relative sliding distance at the interface [7]. de Pannemaecker *et al* [8] proposed an approximation expression of wear rate in random fretting tests. However, for bolted connections, the reasons for changes in normal load under lateral vibrations are complex. Interface wear is just one of these reasons. Therefore, the work rate derived from the above definitions cannot accurately reflect the interface wear process. Consequently, the corresponding wear rate predicted by this work rate is also inaccurate. This is the reason why the work rate is not calculated here [9].

### 4.1.1 Narrowband random process

As mentioned in chapter 2, in the fretting test apparatus the actuator has a steel ball at each end to protect it from shear damage, so it can only push the joint specimen but not pull it back. The pulling-back action is executed by the leaf spring. In addition, the leaf spring also plays a role in limiting the out-of-phase vibration and ensuring the stability of the test apparatus. Thus, the stiffness of the leaf spring is a critical design parameter that controls the output capacity of the piezoelectric actuator. There is a contradiction for the leaf spring between the pulling-back action (requires a large stiffness) and the output capacity (a rigid spring limits the actuator stroke). On the other hand, from the perspective of avoiding resonance, the excitation should be carefully designed to be away from the resonant frequency of the device.

At the same time, even though the actuator is not preloaded, the excitation frequency has a significant effect on the displacement amplitude of the device, because of the limitation of the piezoelectric power amplifier. Therefore, in the random fretting tests only the excitation amplitude is set to be random, while the excitation frequency is considered to be approximately constant. That is, the applied random excitation is a narrowband random process, wherein the signal is distributed over a relatively narrow frequency range.

In general, a wide-band stochastic process can be reconstructed from a given power spectral density (PSD). The root mean square (RMS) of the stochastic signal represents the energy level of the random process. In contrast, a narrowband stochastic process can be expressed as

$$x(t) = a(t)\cos(2\pi f_0 t) - b(t)\sin(2\pi f_0 t), \tag{4.1}$$

where $a$ and $b$ denote the cosine and sine amplitudes (also referred to as low-frequency signal components) of the random process, respectively. They follow a Gaussian distribution law. $t$ is time and $f_0$ is the central frequency of the random process and set to 25 Hz which is the best working frequency of the test device under the comprehensive consideration of the performance and safety of the actuator. Figure 4.1 gives an example of a random process. The low-frequency signal components are generated by low-pass filtering of Gaussian white noise, as depicted

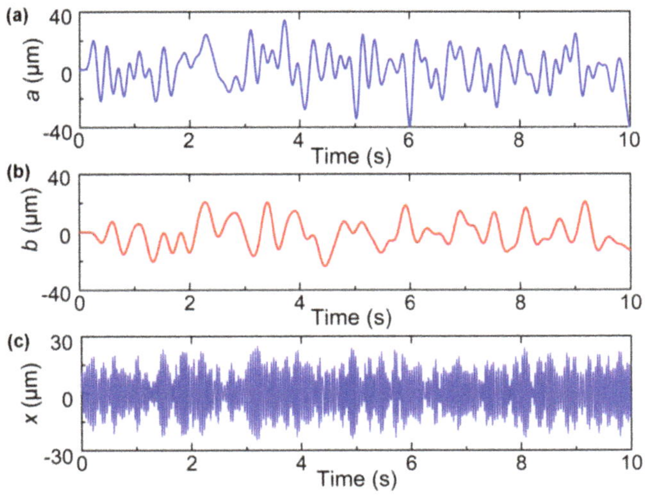


**Figure 4.1.** Zero-mean low-frequency Gaussian signal components of a narrowband stochastic signal (the nominal displacement generated by the piezoelectric actuator): (a) sinusoidal component, (b) cosine component, and (c) synthetic narrowband stochastic signal. (Adapted with permission from [9]. Copyright 2022 Elsevier.)

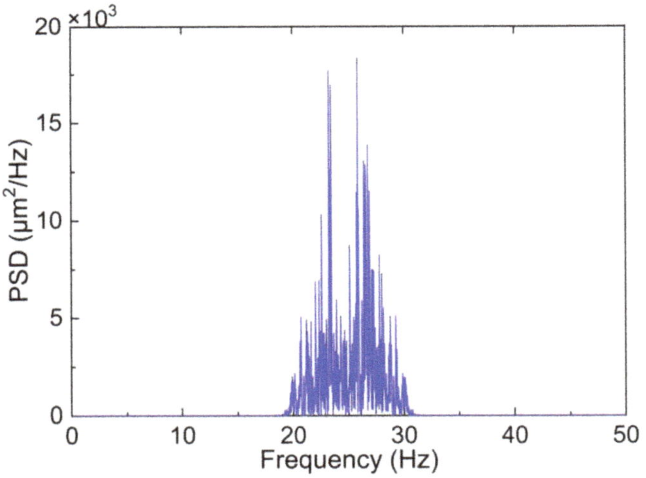


**Figure 4.2.** Power spectral density of the synthetic narrowband stochastic signal. (Reproduced with permission from [9]. Copyright 2022 Elsevier.)

in figures 4.1(a) and (b). Figure 4.2(c) plots the synthesized random signal as a function of time.

Figure 4.2 shows the PSD of the synthesized random signal as a function of frequency. The PSD is distributed in a narrow frequency range (20–30 Hz) centered at 25 Hz.

### 4.1.2 Fretting wear test plan

A set of bolted joint specimens with distinct surface roughness were tested at various excitation levels and preloads. All specimens and bolts were cleaned to remove contaminants and then dried to ensure dry friction conditions. The random fretting tests last 12 h at room temperature. The measured variables including the tangential relative displacement, the friction force, and the bolt preload, were collected using an homemade LabVIEW code. All data were sampled at 5 kHz, and no filtering was applied. To avoid lack of information, a continuous recording strategy rather than interval recording were used.

The analog signal transmitted to the actuator was given by a signal generator. However, due to the limited memory of the signal generator, it is impossible to read a complete 12 h random signal directly through an external device. A 10 s random signal thus was cyclically transmitted to the actuator to perform the 12 h fretting tests. From this viewpoint, the random fretting test here is a stable stochastic process since the statistical characteristics of the excitation signal do not change with time.

The random fretting test plan and operating conditions are listed in table 4.1. The first four tests were carried out under different initial bolt preloads. Bolt pre-tightening is performed with a torque wrench. Once the bolt preload monitored in real-time by the force washer reaches the preset value, the pre-tightening is stopped, such that the initial bolt preloads have a $\pm$ 5% dispersion among different tests.

The influence of transversal excitation levels on interface fretting behavior is shown by comparing tests #2, #5, and #6. Figure 4.3 plots the probability density as a function of the applied displacement excitation (measured by the built-in strain gauge sensor of the actuator) in these three tests. The discrete marked points are calculated by the Matlab function 'ksdensity' [9], while the shaded areas are calculated by the standard Gaussian distribution function, where the mean and variance are obtained from the excitation signal. The probability density of the Gaussian distribution law is expressed as

**Table 4.1.** Summary of the random wear test plan. (Reproduced with permission from [9]. Copyright 2022 Elsevier.)

| Test | RMS of excitation ($\mu$m) | Expectation of excitation ($\mu$m) | Standard deviation ($\mu$m) | Bolt preload, $N$ (kN) $\pm$ 5% dev. | Surface roughness, $R_a$ ($\mu$m) $\pm$ 20% dev. |
|---|---|---|---|---|---|
| #1 | 27.5 | 26.2 | 17.7 | 0.7 | 1 |
| #2 | 27.5 | 26.2 | 17.7 | 1.3 | 1 |
| #3 | 27.5 | 26.2 | 17.7 | 1.8 | 1 |
| #4 | 27.5 | 26.2 | 17.7 | 2.4 | 1 |
| #5 | 22.4 | 21.0 | 14.2 | 1.3 | 1 |
| #6 | 16.6 | 15.7 | 10.8 | 1.3 | 1 |
| #7 | 27.5 | 26.2 | 17.7 | 0.7 | 4 |

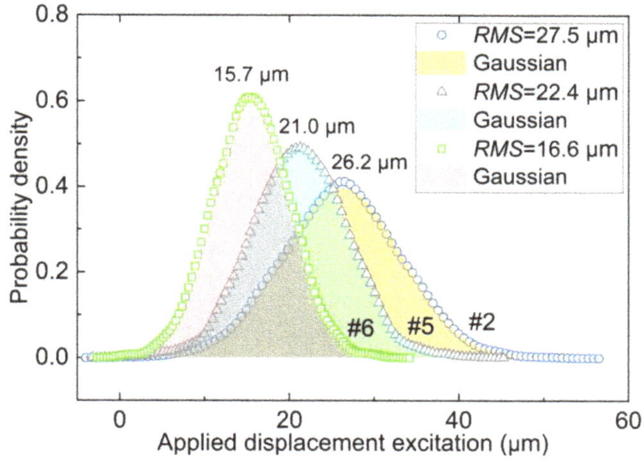


**Figure 4.3.** Probability density as a function of the applied displacement excitation and the corresponding Gaussian distribution for three different cases. (Adapted with permission from [9]. Copyright 2022 Elsevier.)

$$P = \frac{1}{\sqrt{2\pi}\,\sigma} \exp\left[-\frac{(x_i - \mu_x)^2}{2\sigma^2}\right], \tag{4.2}$$

where $\mu_x$ and $\sigma$ denote the expectation and the standard deviation of the distribution, respectively. $x_i$ covers the range of the excitation signal. The discrete marked points are in good agreement with the outlines of the shaded areas, indicating the Gaussian nature of the applied random excitation.

In addition, tests #1 and #7 are carried out and compared to reveal the influence of the surface roughness. It is worth noting that the surface roughness given in table 4.1 is the design value and deviates from its true value by $\pm 20\%$.

### 4.1.3 Stability of random excitation

The excitation system has a built-in closed-loop control module to ensure the stability of the excitation during the entire test. Taking test #1 as an example, the actuator displacements recorded at different times (the first 10 s, after 6 h, and after 12 h) for a duration of 10 s are compared, as shown in figure 4.4. The statistics of these three sets of data are listed in table 4.2, showing good stability of the excitation.

## 4.2 Contact parameter estimation

In harmonic fretting tests, the friction coefficient can be obtained according to the ratio of the friction force at the gross slip regime to the normal force. However, in random fretting tests, the gross slip may not achieved at the joint interface. Therefore, this section introduces a model-based friction coefficient identification method for the case of partial slip.

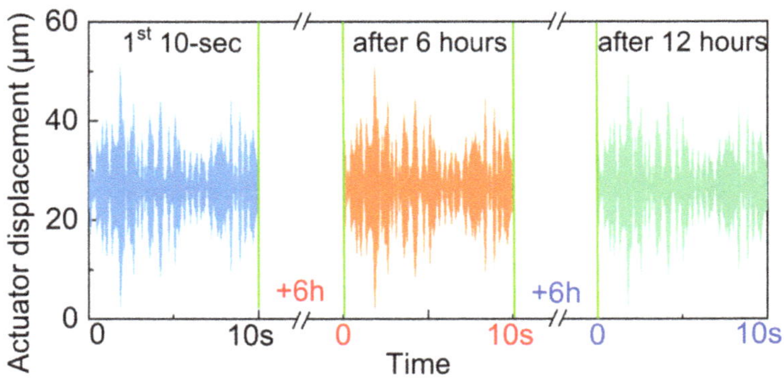


**Figure 4.4.** Actuator displacement measured at different moments (the first 10 s, after 6 h, and after 12 h). (Adapted with permission from [9]. Copyright 2022 Elsevier.)

**Table 4.2.** Statistics of the measured actuator displacements. (Reproduced with permission from [9]. Copyright 2022 Elsevier.)

| Time | Mean ($\mu$m) | Standard deviation ($\mu$m) | Variance ($\mu m^2$) |
|---|---|---|---|
| First 10 s | 26.74 | 6.68 | 44.61 |
| After 6 h | 26.75 | 6.64 | 44.08 |
| After 12 h | 26.75 | 6.64 | 44.06 |

### 4.2.1 Data processing strategy

Figure 4.5 shows the applied displacement excitation $\Delta x$ of the actuator, the tangential relative displacement $\delta$ between the joint interfaces, and the friction force $T$ within the first 10 s in test #1. The interface fretting response shows typical random characteristics. Figure 4.6(a) plots the friction force as a function of the tangential relative displacement in test #1. Unlike the case of sinusoidal tests, the hysteresis loop measured during a loading–unloading cycle in the random test cannot form a closed curve. In some periods, the partial slip and gross slip hysteresis loops may be mixed.

Figure 4.6(b) shows the measured hysteresis loops for the last 10 s in test #1. Comparing with figure 4.6(a), the shape of the hysteresis loop changes significantly with increasing wear cycle. In this case, the interface is always in the partial slip regime. The maximum friction force is larger than that in figure 4.6(a), although the bolt preload is reduced due to interface fretting wear.

In addition, the probability density distribution of the tangential relative displacement hardly changes with fretting wear, as shown in figure 4.7, where the probability density curves at four typical moments are compared. The minor difference among these curves may be due to measurement errors. These curves still follow a Gaussian distribution as the wear cycle increases.

To estimate the contact parameters (mainly the friction coefficient and tangential contact stiffness), the measured relative displacement signal is divided into several

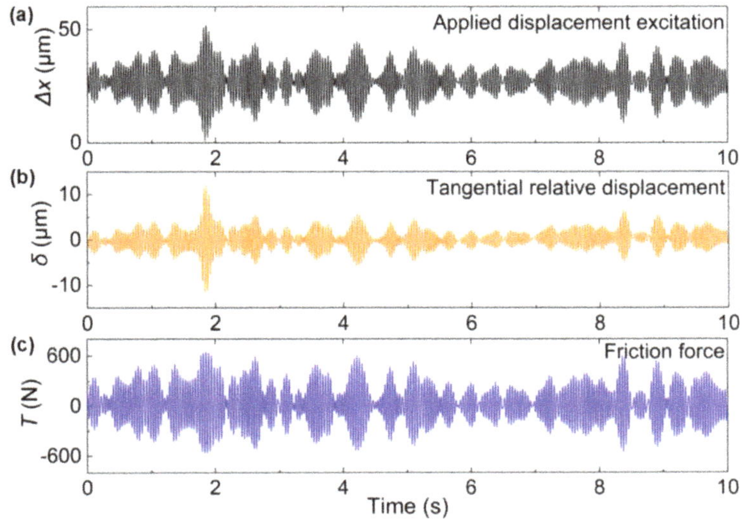


**Figure 4.5.** Measured time–history curves within the first 10 s in test #1: (a) applied displacement excitation measured by the built-in strain gauge sensor of the actuator, $\Delta x$, (b) tangential relative displacement, $\delta$, and (c) friction force, $T$. (Adapted with permission from [9]. Copyright 2022 Elsevier.)

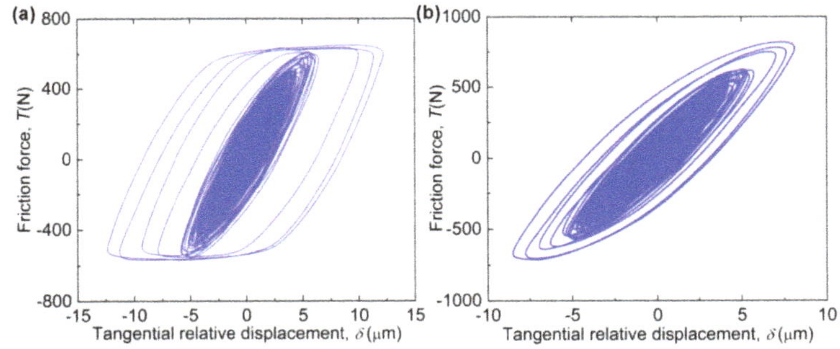


**Figure 4.6.** Measured friction force as a function of tangential relative displacement in test #1: (a) for the first 10 s and (b) for the last 10 s. (Reproduced with permission from [9]. Copyright 2022 Elsevier.)

segments using adjacent peak points as breakpoints, as illustrated in figure 4.8, along with the friction force. Each segmented displacement and friction force can form a non-closed hysteresis loop from which the contact parameters can be estimated. Given that the contact parameters hardly change over a short period, for simplicity, the following signal division strategy is used for contact parameter estimation: for the first 20 min, the length of each divided signal segment is 1 s (that is, there are 1200 sets of signals with 1 s time length); for the rest, the length of each divided signal segment is 10 s.

Then, from each segment of the signal, the hysteresis loop with the largest displacement amplitude is selected to estimate the corresponding contact parameters. The reason for selecting the maximum hysteresis loop is that, in terms of computational cost, the conventional method for estimating contact parameters is

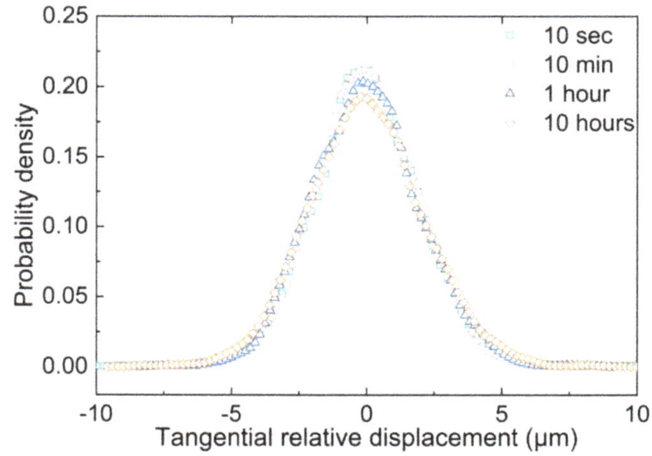


**Figure 4.7.** Effect of loading duration on the probability density of the tangential relative displacement. (Reproduced with permission from [9]. Copyright 2022 Elsevier.)

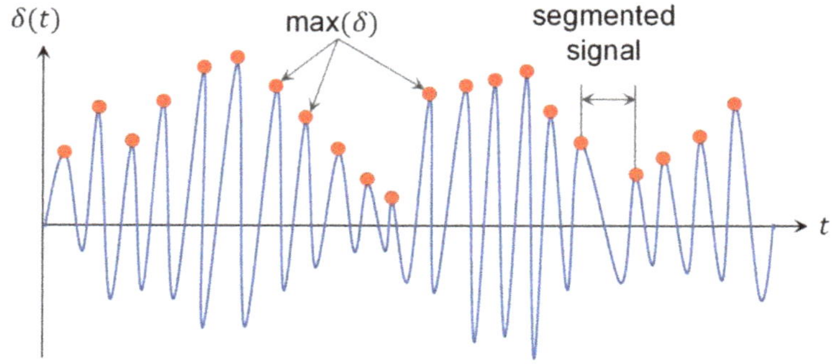


**Figure 4.8.** Displacement signal segmentation for estimating contact parameters. (Reproduced with permission from [9]. Copyright 2022 Elsevier.)

more effective than the model-based method in the case of gross slip. Details will be given in the next section.

### 4.2.2 Contact parameter estimation method

As shown in chapter 3, tangential contact stiffness $k_t$ and friction coefficient $\mu$ of bolted joint interfaces gradually evolve with increasing wear. In random fretting tests, high-amplitude (gross slip) and low-amplitude (partial slip) hysteresis loops are mixed. Although the measured signal is split into several segments and the hysteresis loop with the largest relative displacement amplitude is chosen to extract contact parameters, the gross slip regime does not always occur. Therefore, we require two different parameter estimation methods for high-amplitude and low-amplitude hysteresis loops. Note that the estimations of the tangential contact stiffness are the same in both cases.

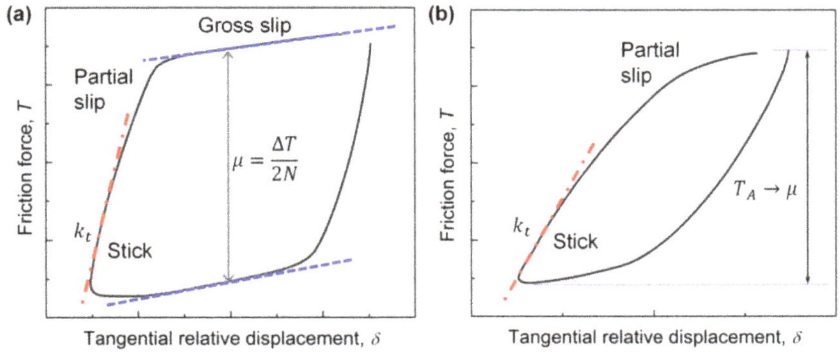


**Figure 4.9.** Effect of loading duration on the probability density of the tangential relative displacement. (Reproduced with permission from [9]. Copyright 2022 Elsevier.)

The tangential contact stiffness is defined as the slope of the force–displacement curve at the stick stage, i.e. $k_t = \Delta T_{\text{stick}}/\Delta\delta_{\text{stick}}$, as shown in figure 4.9 (see the dash-dotted line). The increments $\Delta T_{\text{stick}}$ and $\Delta\delta_{\text{stick}}$ are defined choosing 20 points within the linear portion of the loop at the reversal. For high-amplitude hysteresis loops, the friction coefficient is usually determined by the ratio between the friction force in the gross slip regime and the bolt preload. However, due to the bending of the bolt shank, the friction force in the gross slip regime is not constant, see the dashed line in figure 4.9(a). Therefore, the friction coefficient is defined as the ratio of the friction difference to twice the bolt preload [11],

$$\mu = \frac{\Delta T}{2N}, \tag{4.3}$$

where $\Delta T$ is the difference between the friction forces during the loading and unloading gross slip regime, $N$ the bolt preload. Alternatively, based on energy loss, it can be approximately expressed as follows [12]:

$$\mu \approx \frac{E_{\text{loop}}}{4N\delta_A}, \tag{4.4}$$

where $E_{\text{loop}}$ denotes the energy loss per cycle (i.e. the area enclosed by hysteresis loop) and $\delta_A$ the amplitude of the tangential relative displacement.

For low-amplitude hysteresis loops (see figure 4.9(b)), since the interface motion does not induce a gross slip condition, it is only known that the friction coefficient should be greater than the maximum ratio of friction force to bolt preload, $\mu > T_A/N$, where $T_A$ is the amplitude of the friction force.

To accurately estimate the friction coefficient in the partial slip case, a model-based approach is used. The study in [13] demonstrates through a simulation–experiment comparison that the friction coefficient significantly influences the range of friction force in the hysteresis loop. This is a typical single-objective optimization problem, in which the objective function and the range of the design variable are

$$\min\{\Delta T_{\text{exp}} - \Delta T_{\text{sim}}(\mu)\}, \quad \frac{T_A}{N} < \mu < 1, \tag{4.5}$$

where $\Delta T_{\text{exp}}$ and $\Delta T_{\text{sim}}$ represent the difference between the upper and lower limits of the friction force in the experimental and simulated hysteresis loops, respectively. There are a variety of contact models [14–16] available for implementing friction hysteresis simulations. Considering the simplicity of model parameters, the Iwan model is employed to reproduce the friction hysteresis behavior. The original Iwan model can be characterized by three parameters: tangential contact stiffness, friction coefficient and normal preload. A detailed description of the Iwan model will be given in the next chapter.

Figures 4.10–4.12 provide an example of friction coefficient estimation and validate the effectiveness of this method through cross-comparison. The exemplary experiment is conducted under sinusoidal vibration and a bolt preload of 2180 N. Unlike typical sinusoidal tests, the excitation signal used here is composed of two sinusoidal signals at 25 Hz with different amplitudes, sequentially concatenated. The low-amplitude excitation does not cause gross slip at the interface, whereas high-amplitude excitation induces the gross slip. The model-based method (equation (4.5)) and the traditional method (equation (4.3)) are used to estimate the friction coefficients of these two cases, respectively. Since the test time is very short, only 0.14 s, the effect of wear on the friction coefficient can be ignored, such that the friction coefficients obtained by the two methods should be the same.

Figure 4.10 plots the measured tangential relative displacement and friction force as a function of time. It can be seen from the friction force curve that the interface partial slip occurs in the A–B stage, while the gross slip occurs in the B–C stage. This is more evident in figure 4.11 which depicts the hysteresis loops during the A–B and B–C stages.

According to the recognition method of the tangential contact stiffness, it is easy to get its value from the hysteresis loops, $k_t = 211$ N $\mu$m$^{-1}$. In the partial slip and gross slip situations, this value remains unchanged. Figure 4.12 compares the

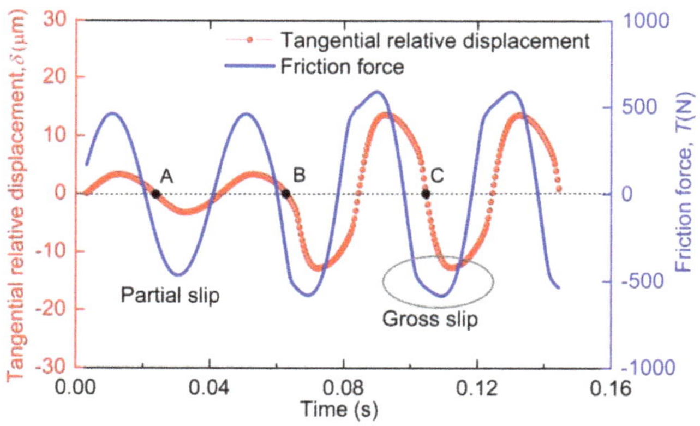


**Figure 4.10.** Tangential relative displacement and friction force as a function of time under sinusoidal vibrations with different amplitudes. (Reproduced with permission from [9]. Copyright 2022 Elsevier.)

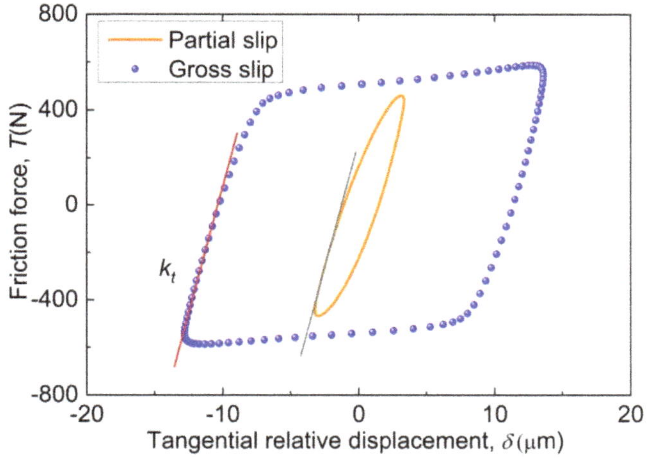


**Figure 4.11.** Two adjacent hysteresis loops (continuously measured, such as the time history from points A to C in figure 4.10) with different excitation amplitudes. (Reproduced with permission from [9]. Copyright 2022 Elsevier.)

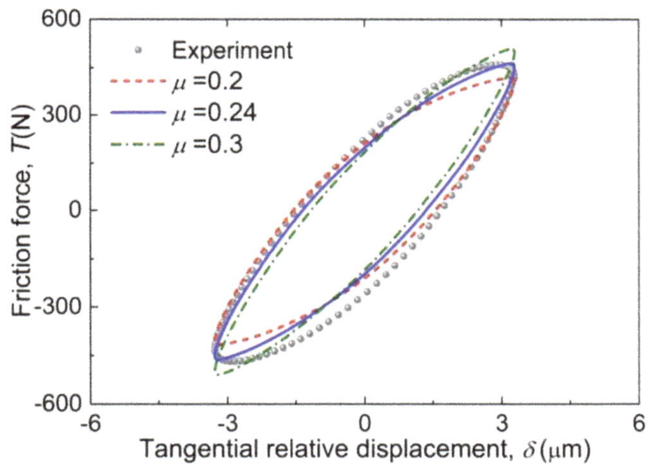


**Figure 4.12.** The identification process of friction coefficients under the interface partial slip regime by adjusting the preset friction coefficient to match the predicted friction amplitude with the corresponding experimental value. (Reproduced with permission from [9]. Copyright 2022 Elsevier.)

measured hysteresis loop and the simulated ones in which different friction coefficients are considered. The simulation results show the friction force amplitude varies monotonically with the friction coefficient. From the matching degree of the friction force range in the hysteresis loop, the friction coefficient is estimated to be 0.24. This value is consistent with that estimated from the high-amplitude hysteresis loop by equation (4.3). Therefore, the feasibility of the model-based method is demonstrated.

## 4.3 Results and discussion

### 4.3.1 Effect of bolt preload

Four joint specimens (in the first four tests) are preloaded with different initial loads (0.7, 1.3, 1.8, and 2.4 kN) and tested under the same transversal excitation level (RMS = 27.5 $\mu$m). Figure 4.13 shows the variation of bolt preload with test time, where $N_0$ is the initial value of bolt preload $N$. In all four test groups, the bolt preload decreases to varying degrees immediately after loading. This decline occurs at the moment of loading and the drop is about 5%–8% of the initial value. Subsequently, the bolt preload gradually decreases and eventually tends to be asymptotically steady. The smaller the initial preload, the more pronounced the second decay of the bolt preload (after the initial rapid decline).

The causes of bolt loosening are varied [17, 18]. In the above tests, on one hand, material removal due to interface wear reduces the interference fit between the mating surfaces, leading to a decrease in bolt preload. On the other hand, the transversal vibration of joint specimens cause bending deformation of the bolt shank, resulting in fretting wear between the mating threads, which also contributes to the loosening of bolt preload.

Figure 4.14 shows the variation trend of the friction coefficient with cumulative energy loss. The cumulative energy loss $E_c$ is calculated by the following equation

$$E_c = \int_0^t \int_{\delta_{\min}}^{\delta_{\max}} T(\delta)\mathrm{d}\delta\mathrm{d}t, \tag{4.6}$$

where $t$ denotes time, $T$ is friction force, $\delta_{\max}$ and $\delta_{\min}$ are the maximum and minimum values of relative displacement. The results show that the variation of the friction coefficient in these four tests is similar to the trend observed in conventional wear tests [19–21]. The friction coefficient increases rapidly at the beginning of wear and then tends to stabilize. The rapid and significant increase in friction coefficient at

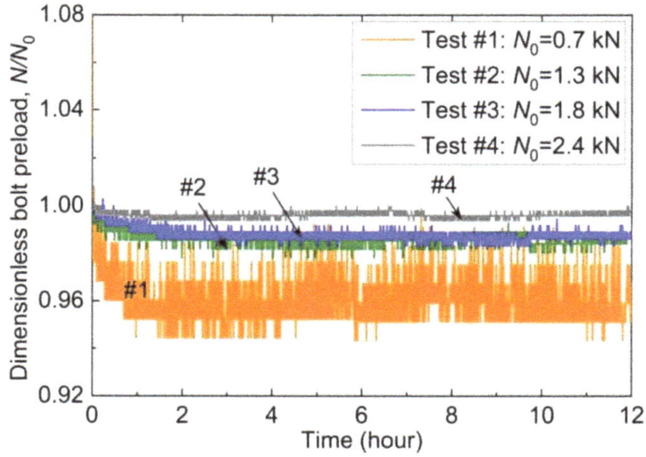


**Figure 4.13.** Dimensionless bolt preload $N/N_0$ ($N_0$ denotes the initial value of the bolt preload $N$) as a function of time in the first four tests. (Reproduced with permission from [9]. Copyright 2022 Elsevier.)

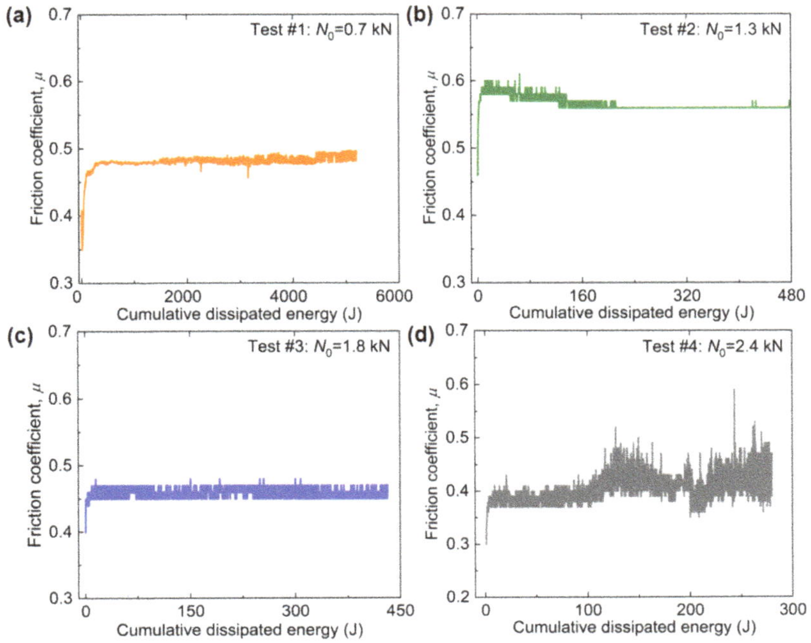


**Figure 4.14.** Evolution of friction coefficients with cumulative dissipated energy in the first four tests. (Adapted with permission from [9]. Copyright 2022 Elsevier.)

the initial stage is due to the generation of wear particles on the contact surface and rapid increase in the number of particles. The subsequent steady state may be due to the balance between the generation and expulsion of wear particles at the interface. Although the surface roughness measured in these four tests is close, the initial friction coefficients show significant variability, which may be caused by the surface waviness.

The influence of bolt preload on the results is reflected in two aspects. First, the greater the initial preload, the less the final cumulative dissipated energy. Second, the smaller the initial preload, the more pronounced the increase in the friction coefficient during the initial wear phase. This is because higher bolt preload limits the occurrence of the gross slip regime.

Figure 4.15 shows the variation of tangential contact stiffness with cumulative energy dissipation. Different from traditional sinusoidal wear tests, in these four tests, the tangential contact stiffness fluctuated within a small range of $\pm 10$ N $\mu$m$^{-1}$ around a stable value. Generally speaking, there are many factors that affect the tangential contact stiffness [22–24]. In addition to the material properties, the dominate factors are normal load, surface roughness and waviness, and sliding velocity. The normal load affects the deformation degree and actual contact area of the contact area, thereby changing the tangential contact stiffness. The influence of surface roughness and waviness on tangential contact stiffness is reflected in the actual contact area of the contact region. Sliding velocity may affect the temperature

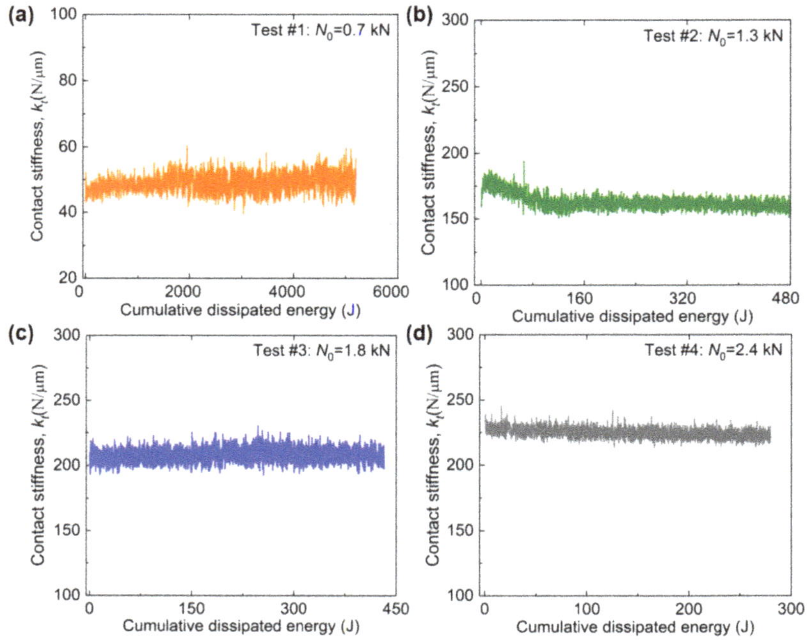


**Figure 4.15.** Evolution of tangential contact stiffness with cumulative dissipated energy in the first four tests. (Adapted with permission from [9]. Copyright 2022 Elsevier.)

and contact properties of the friction interface, thus influencing the tangential contact stiffness.

As mentioned in chapter 3, the tangential contact stiffness is directly related to the actual contact area. In tests #1–#4, the actual contact area is related to the wear process and the bolt preload. The wear process increases the contact area, leading to an increase in tangential contact stiffness. At the same time, a decrease in preload reduces the contact area, resulting in a drop in tangential contact stiffness. The phenomenon shown in figure 4.15, where greater preload leads to larger contact stiffness, also indirectly supports this statement.

### 4.3.2 Effect of excitation level

Bolted joint specimens were tested at the same initial preload and different excitation levels to observe the effect of excitation levels. Figure 4.16 shows the bolt preload as a function of test time. Similar to previous results, in these three tests, the bolt preload decreases at initial loading phase. Apart from this decrease, the preload remains almost constant as wear increased.

Figures 4.17(a)–(c) plot the friction coefficient versus the cumulative energy dissipation for different excitation levels. The final cumulative energy loss in test #2 is more than twice that dissipated in tests #5 and #6 due to the different severity of fretting wear. The friction coefficient increases rapidly at the early stage and then reaches a stable state. As mentioned above, the evolution of friction coefficients is

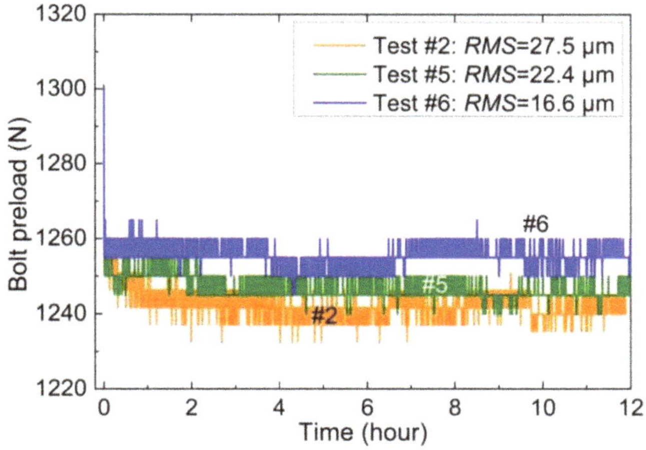


**Figure 4.16.** Evolution of bolt preloads with time in tests #2, #5, and #6. (Reproduced with permission from [9]. Copyright 2022 Elsevier.)

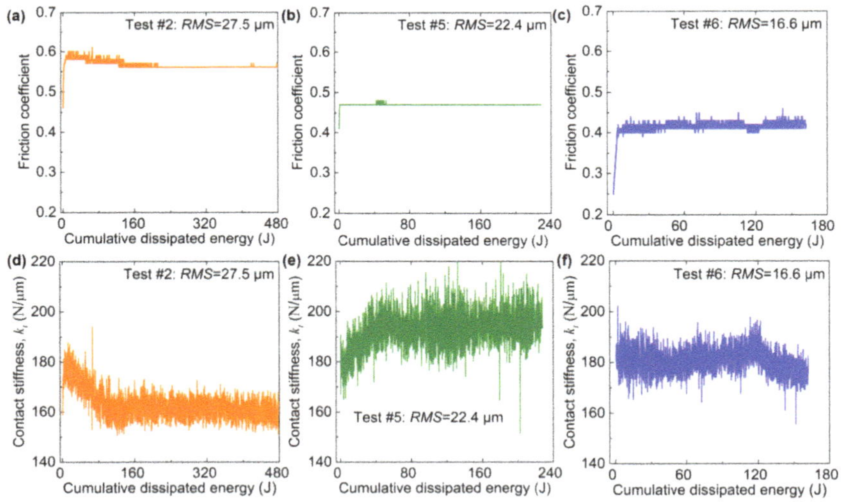


**Figure 4.17.** Evolution of contact parameters with cumulative dissipated energy: (a)–(c) friction coefficients, and (d)–(f) tangential contact stiffness in tests #2, #5, and #6. (Reproduced with permission from [9]. Copyright 2022 Elsevier.)

mainly controlled by the generation and discharge of wear particles at the interface. Especially in test #2, the friction coefficient shows a peak before the steady-state stage, which may be caused by the strong local plastic deformation of the contact surface. Figures 4.17(d)–(f) show the variation trend of the tangential contact stiffness with cumulative energy dissipated. In test #2, the tangential contact stiffness initially increases rapidly, then exhibits significant fluctuations and decreases before finally reaching a steady state. The initial increase is due to the wear-induced increase in contact area, while the subsequent decrease was caused by the preload reduction-induced decrease in contact area. In test #5, the tangential contact stiffness increases slowly and then tends to be stable; in test #6, the tangential contact stiffness is always in a dynamically stable state.

### 4.3.3 Effect of surface roughness

Bolted joint interfaces with different surface roughness ($R_a = 4\ \mu$m in test #7 and $R_a = 1\ \mu$m in test #1) are compared to reveal the effect of surface roughness. Figure 4.18 depicts the bolt preload versus test time. Overall, the differences among the three are significant, similar to the results of the sinusoidal vibration tests in chapter 3. The reduction in preload for the rough surface (test #7) is more pronounced than for the smooth surface (test #1). In test #7, the final reduction even reached 35% of the initial value. For surfaces with higher roughness, the asperities of the rough surface are more easily cut and flattened during the wear process. Therefore, the attenuation of the interference fit between the contact surfaces is more evident.

Figure 4.19 shows the friction coefficient and tangential contact stiffness as a function of cumulative energy loss. It can be seen that the rough surface ends up dissipating slightly more energy than the smooth surface. Throughout the entire test cycle, the friction coefficient of the rough surface was greater than that of the smooth surface, and the same applies to the tangential contact stiffness. In these two tests, the friction coefficient increases rapidly and significantly during the initial wear phase. Subsequently, in test #1, it remained stable, whereas in test #7, the friction coefficient undergoes slow oscillations before reaching a final steady state. This oscillation may be caused by the grinding of large wear particles generated on the rough surface.

The tangential contact stiffness of the rough surface is even three times that of the smooth surface, indicating that the actual contact area of the former is much higher

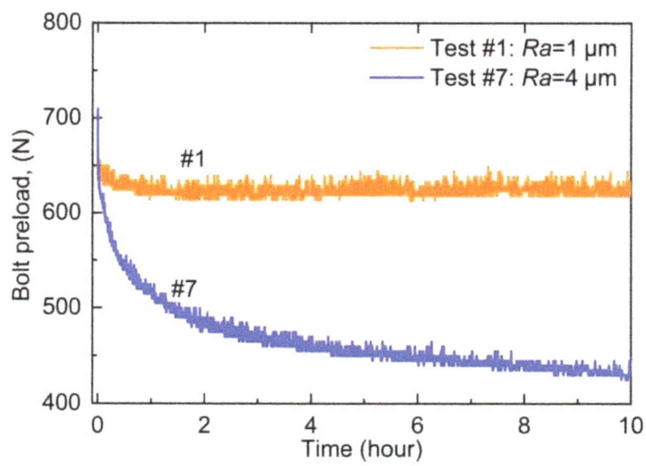


**Figure 4.18.** Evolution of bolt preloads with time in tests #1 and #7. (Reproduced with permission from [9]. Copyright 2022 Elsevier.)

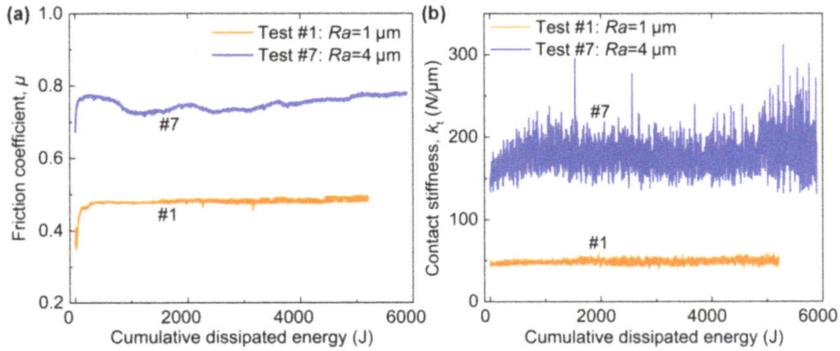


**Figure 4.19.** Contact parameters as a function of cumulative dissipated energy: (a) friction coefficients and (b) tangential contact stiffness in tests #1 and #7. (Reproduced with permission from [9]. Copyright 2022 Elsevier.)

than that of the latter. In test #7, although the bolt preload decreases significantly, the tangential contact stiffness does not decrease but remained in a dynamically stable state. This is because the decrease in contact area caused by the preload reduction is offset by the increase in contact area due to wear.

## 4.4 Summary

This chapter introduces a fretting wear test method on the bolted joint interface under transversal random vibrations and the interface fretting response. In random fretting tests, it is necessary to ensure the stability of the excitation. This is typically assessed by monitoring the changes in the expectation and variance of the random excitation over time. In addition, when analyzing the measured data, it is essential to ensure the accuracy of the identified contact parameters as much as possible. This chapter introduces a model-based method for identifying the friction coefficient specifically for partial slip conditions. The main conclusions are as follows:

1. The friction coefficient initially increases rapidly and significantly with wear, then gradually stabilizes, regardless of changes in initial bolt preload and excitation level. However, a lower initial preload results in a more pronounced change in the friction coefficient, due to reduced interface gross slip at higher preloads.
2. The tangential contact stiffness remained approximately stable but exhibited noticeable oscillations. This behavior is attributed to the relatively low excitation level, which makes the evolution of the wear scar (actual contact area) less pronounced, although the contact surface is in the mixed-fretting regime under certain conditions. The rough surface shows more drastic changes in bolt preload and contact stiffness compared to the smooth surface.

Parts of this chapter have been reproduced with permission from [9].

## Bibliography


[1] de Pannemaecker A, Fouvry S, Brochu M and Buffiere J Y 2016 Identification of the fatigue stress intensity factor threshold for different load ratios *R*: from fretting fatigue to C(T) fatigue experiments *Int. J. Fatigue* **82** 211–25

[2] Attia H, Meshreki M, Korashy A, Thomson V and Chung V 2011 Fretting wear characteristics of cold gas-dynamic sprayed aluminum alloys *Tribol. Int.* **44** 1407–16

[3] Heredia S and Fouvry S 2010 Introduction of a new sliding regime criterion to quantify partial, mixed and gross slip fretting regimes: correlation with wear and cracking processes *Wear* **269** 515–24

[4] Zhou Z R and Vincent L 1995 Mixed fretting regime *Wear* **181** 531–6

[5] de Pannemaecker A, Attia H and Williams G 2019 A novel acceleration-controlled random vibration fretting test methodology: from classical sinusoidal to Gaussian random excitation *Wear* **438** 203050

[6] Zhu M H and Zhou Z R 2011 On the mechanisms of various fretting wear modes *Tribol. Int.* **44** 1378–88

[7] Attia H 2009 A generalized fretting wear theory *Tribol. Int.* **42** 1380–8

[8] de Pannemaecker A, Attia H and Williams G 2021 Characterization of the fretting wear damage under random excitation *Wear* **476** 203637

[9] Li D, Xu C, Li R and Zhang W 2022 Contact parameters evolution of bolted joint interface under transversal random vibrations *Wear* **500** 204351

[10] ksdensity *MathWorks* https://mathworks.com/help/stats/ksdensity.html?s_tid=srchtitle_ksdensity_1

[11] Schwingshackl C W, Petrov E P and Ewins D J 2012 Measured and estimated friction interface parameters in a nonlinear dynamic analysis *Mech. Syst. Sig. Process.* **28** 574–84

[12] Lavella M, Botto D and Gola M M 2013 Design of a high-precision, flat-on-flat fretting test apparatus with high temperature capability *Wear* **302** 1073–81

[13] Li D, Xu C, Liu T, Gola M M and Wen L 2019 A modified IWAN model for micro-slip in the context of dampers for turbine blade dynamics *Mech. Syst. Sig. Process.* **121** 14–30

[14] Segalman D J 2005 A four-parameter Iwan model for lap-type joints *J. Appl. Mech.* **72** 752–60

[15] Ikhouane F and Rodellar J 2005 On the hysteretic Bouc–Wen model *Nonlinear Dyn.* **42** 79–95

[16] Freidovich L, Robertsson A, Shiriaev A and Johansson R 2009 LuGre-model-based friction compensation *IEEE Trans. Control Syst. Technol.* **18** 194–200

[17] Huang J, Liu J, Gong H and Deng X 2022 A comprehensive review of loosening detection methods for threaded fasteners *Mech. Syst. Sig. Process.* **168** 108652

[18] Miao R, Shen R, Zhang S and Xue S 2020 A review of bolt tightening force measurement and loosening detection *Sensors* **20** 3165

[19] Fantetti A, Tamatam L R, Volvert M, Liu I, Salles L, Brake M R W, Schwingshackl C W and Nowell D 2019 The impact of fretting wear on structural dynamics: experiment and simulation *Tribol. Int.* **138** 111–24

[20] Yoon Y, Etsion I and Talke F E 2011 The evolution of fretting wear in a micro-spherical contact *Wear* **270** 567–75

[21] Hintikka J, Lehtovaara A and Mäntylä A 2015 Fretting-induced friction and wear in large flat-on-flat contact with quenched and tempered steel *Tribol. Int.* **92** 191–202

[22] Kartal M E, Mulvihill D M, Nowell D and Hills D A 2011 Measurements of pressure and area dependent tangential contact stiffness between rough surfaces using digital image correlation *Tribol. Int.* **44** 1188–98
[23] Yang H, Che X and Yang C 2020 Investigation of normal and tangential contact stiffness considering surface asperity interaction *Ind. Lubr. Tribol.* **72** 379–88
[24] Fuadi Z, Takagi T, Miki H and Adachi K 2013 An experimental method for tangential contact stiffness evaluation of contact interfaces with controlled contact asperities *Proc. Inst. Mech. Eng.* J **227** 1117–28

# Part II

## Modeling friction and wear

**IOP** Publishing



**Chao Xu and Dongwu Li**

# Chapter 5

# A phenomenological modeling method for friction hysteresis

An accurate description of friction hysteresis is a prerequisite for successfully predicting the dynamics of joint structures. This chapter introduces a generalized friction model in the framework of the Iwan model, which can better simulate the nonlinear constitutive relationship at joints. This method obtains the Iwan density function from the contact pressure distribution on joint interfaces, without having to assume it like traditional Iwan-type models. A detailed derivation of the interface force–displacement expression under monotonic and cyclic loading is given. Moreover, the generalized Iwan model is applied to two typical contact geometries: sphere-on-sphere and flat-on-flat.

## 5.1 Introduction

Phenomenological models are widely used to reproduce the nonlinear hysteretic friction behavior of connection interfaces due to their advantages such as simple expressions, few parameters, and easy integration into finite element codes. Over the past few decades, researchers have developed several phenomenological friction models, such as the Iwan model [1, 2], the Valanis model [3–5], and the Bouc–Wen model [6–9]. The Iwan and Valanis models were initially proposed to reproduce the elastoplastic behavior of metallic materials and was later used to simulate the frictional hysteresis behavior of joint interfaces. The Bouc–Wen model is a mathematical model widely used to describe hysteresis in systems subjected to cyclic loading in structural and material engineering. It is defined by a differential equation that includes parameters controlling the shape and size of the hysteresis loop, allowing it to simulate various types of nonlinear hysteresis phenomena. Of course, there are other types of friction models [10–12].

Phenomenological friction models should meet several important requirements to be effective: (i) the model must be capable of describing multiple friction states (i.e.

doi:10.1088/978-0-7503-6214-6ch5  

stick, partial slip, and gross slip) and reproducing the amplitude-dependent frictional damping; (ii) model parameters should be easy to estimate; and (iii) the model should be compatible with finite element codes [13].

The Iwan model is frequently used to analyze the mechanics and dynamics of joint structures due to its ability to describe typical friction phenomena and the ease with which its parameters can be extracted. There are many modifications of the Iwan model [14–25]. These Iwan-type models can be divided into two categories based on different modeling strategies. The first approach combines the framework of the Iwan model with tribological methods based on rough contact theory, which can be regarded as physics-based. However, its force–displacement formula is complex and may not even have an explicit expression.

The second scheme is based on the essence of the original Iwan model and aims to improve accuracy and completeness. Segalman [26] developed a four-parameter Iwan model that accounts for the power-law relationship between energy dissipation and the amplitude of the applied load. He assumed the Iwan density function to be a truncated power-law distribution with one Dirac delta function. Wang *et al* [18] proposed an improved Iwan model based on the four-parameter Iwan model to describe the residual stiffness during large slip processes and the smooth transition of connection stiffness from partial slip to gross slip conditions. Furthermore, Li *et al* [19, 20] presented a six-parameter Iwan model to simulate the friction at lap joints using a truncated power-law distribution with two Dirac delta functions. This model considers the residual stiffness during the gross slip regime and the power-law relationship of energy dissipation in the partial slip regime. An often overlooked fact about bolted joints is the contact between the bolt shank and the bolt hole under transversal loads due to misalignment, as shown in chapter 3. Of course, this contact may also be caused by excessive transversal loads causing the sliding distance between the contact interfaces to exceed the distance between the bolt shank and the hole. Brake [21] extended the four-parameter Iwan model and developed an improvement Iwan model, which can consider the 'pinning' effect between the screw and the bolt hole. Similar works can also be found in [27, 28]. Different from the abovementioned models, Rajaei *et al* [23] and Li *et al* [24, 25] considered the effect of normal load variation on interface friction hysteresis. They proposed a modified Iwan model representing both normal load variations and tangential stick–slip behavior.

Although the aforementioned Iwan-type models can describe interface phenomena observed in some experiments, a fundamental fact, namely the influence of normal load distribution on tangential friction behavior, is rarely considered. In essence, the contact pressure distribution determines the motion state of local regions on the contact surface, whether it is sticking or sliding [29]. In practical applications, the contact pressure distribution is influenced by the geometry of contacting bodies and surface roughness. This may be the main reason for the observed differences in the power-law relationship between energy dissipation and the amplitude of applied force in some measurements [30]. Li *et al* [13] established the relationship between contact pressure distribution function and the Iwan density function and proposed a generalized Iwan model to further better describe the friction hysteresis behavior at joint surfaces. This chapter focuses on the theoretical framework of the generalized Iwan model and its application in typical contact problems.

## 5.2 Generalized Iwan model

This section briefly reviews the Iwan model and emphasizes the importance of the Iwan density function (IDF). Then, the generalized Iwan model is explained in detail to obtain the IDF from the pressure distribution at contact surfaces. This approach provides a physical interpretation of the IDF and does not introduce new model parameters.

### 5.2.1 The Iwan model and density function

The Iwan model [1] consists of a series of Jenkins elements in parallel, as shown in figure 5.1(a). The Jenkins element is a bilinear model that can be used to characterize the stick and slip states at contact interfaces [31], as shown in figure 5.2. It is composed of a linear spring with stiffness $k_t/n$ and a Coulomb slider with a critical sliding force $f_i^*/n$, where $k_t$ denotes the tangential contact stiffness of the overall interface, $f_i^*$ is the critical sliding force on the $i$th element, and $n$ is the number of the Jenkins element. The critical sliding force here refers to the yield force of the Coulomb slider, that is, the force acting on the Jenkins element when it starts to slide.

It should be noted that the critical sliding force of each element is different and can be distributed according to the IDF $\varphi(f^*)$. $\varphi(f^*)\mathrm{d}f^*$ is the fraction of the total

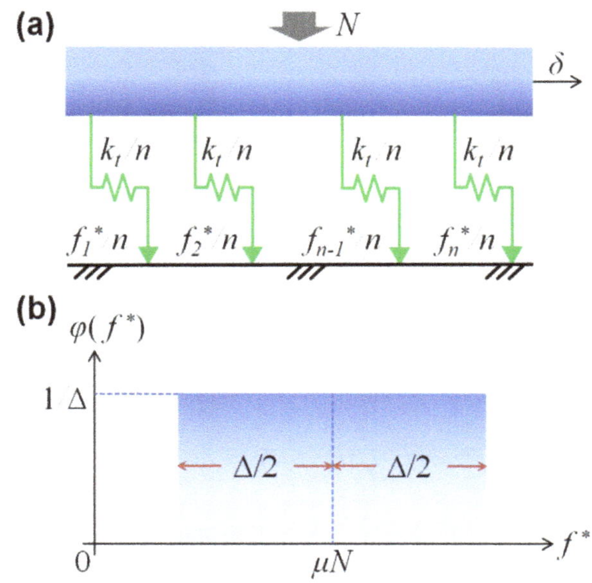


**Figure 5.1.** (a) Schematic diagram of the Iwan model and (b) the density function of critical sliding force assumed in the original Iwan model.

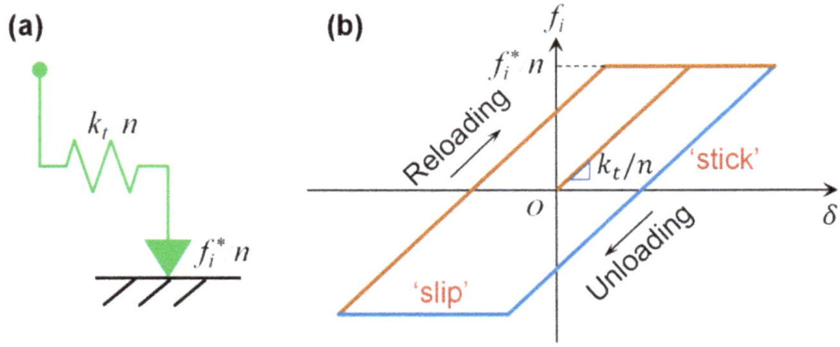


**Figure 5.2.** (a) Jenkins element and (b) hysteresis loop under initial monotonic loading and cyclic loading.

number of Jenkins elements having $f^* \leqslant f_i^* \leqslant f^* + \mathrm{d}f^*$ [1]. The total critical sliding force of the Iwan model is equal to the Coulomb friction force $\mu N$ at contact surfaces, where $\mu$ is the friction coefficient and $N$ the normal load. The original Iwan model assumes an uniform IDF, as shown in figure 5.1(b),

$$\varphi(f^*) = \frac{1}{\Delta}, \tag{5.1}$$

where $\Delta$ is the range of $f^*$.

Under a certain load, the Jenkins element in the Iwan model may be in a stick state or a slip state. In the stick state, $k_t\delta/n < f_i^*/n$ where $\delta$ denotes the relative displacement at contact interfaces, while in the slip state, $k_t\delta/n \geqslant f_i^*/n$. Therefore, for a small-amplitude monotonic loading case, the total friction force can be expressed as

$$T(\delta) = \int_0^{k_t\delta} f^*\varphi(f^*)\mathrm{d}f^* + k_t\delta \int_{k_t\delta}^{\infty} \varphi(f^*)\,\mathrm{d}f^*. \tag{5.2}$$

The first integral term represents the force on all Jenkins elements in the slip state, while the second integral term represents the force on all Jenkins elements in the stick state. Once the friction force exceeds the critical sliding force of the Iwan model, all Jenkins elements are in a slip state, and the total friction force is reset to $\mu N$.

For a cyclic load, the force–displacement relationship of the Iwan model during unloading can be derived according to the Masing hypothesis [16, 26] as follows,

$$T(\delta) = \int_0^{\frac{k_t(\delta_m-\delta)}{2}} -f^*\varphi(f^*)\mathrm{d}f^* + \int_{\frac{k_t(\delta_m-\delta)}{2}}^{k_t\delta_m} [f^* - k_t(\delta_m - \delta)]\varphi(f^*)\mathrm{d}f^* + k_t\delta \int_{k_t\delta_m}^{\infty} \varphi(f^*)\mathrm{d}f^*, \tag{5.3}$$

where $\delta_m$ represents the amplitude of tangential relative displacement. Similarly, the expression of the friction force during reloading can also be deduced in the same way.

There are three parameters (namely, tangential contact stiffness $k_t$, total critical sliding force $\mu N$, and the range of the critical sliding force) in the original Iwan model for describing the friction–displacement relationship, among which the ratio of the critical sliding force range over the critical force is related to the IDF. This ratio can be obtained by fitting with experimental results. It has significant impact on the description of the partial slip state and alters the shape of hysteresis loops. Therefore, it can be seen that the IDF plays a key role in the Iwan model. The next section will introduce a generalized Iwan model in which the IDF can be explicitly derived from contact pressure distribution on joint surfaces.

### 5.2.2 Modeling method of generalized Iwan model

Subjected to a tangential load, the contact surface can be divided into two regions in accordance with the contact state: one is the slip region, where relative motion occurs between the contact pairs; the other is the stick region in which the interface relative motion is not allowed and the deformation at the interface is mainly elastic. For example, in the partial slip regime of a spherical contact, the slip region is the annular periphery of the contact area, where the contact pressure is relatively lower and cannot constrain the relative motion between contact pairs. Therefore, it is evident that the motion state of contact surfaces significantly depends on the contact pressure distribution. In view of this, Li *et al* [13] related the IDF of the critical sliding force to the contact pressure distribution and gave the IDF an explicit physical significance.

Figure 5.3 shows the flow chart of the modeling approach of the generalized Iwan model and illustrates it using the original Iwan model as an example. Step 1 is to

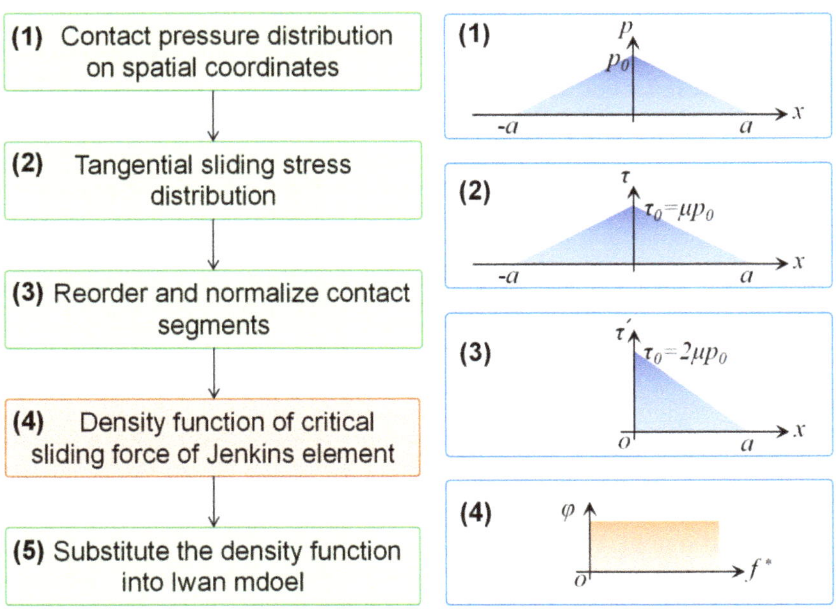


**Figure 5.3.** Flow chart of the modeling approach of generalized Iwan model and an example of a 'triangle' pressure distribution. (Reproduced with permission from [13]. Copyright 2020 Elsevier.)

obtain the contact pressure distribution function at interfaces from analytical or semi-analytical solutions, measurements, or finite element analysis, as follows,

$$p = p(N, x, a), \ |x| \leqslant a, \tag{5.4}$$

where $x$ and $a$ denote spatial coordinate axis and the radius of contact area, respectively. Throughout the modeling process, the distribution function of the contact pressure is assumed to be independent of the relative motion of the interface.

Step 2 establishes the distribution function of the tangential sliding stress $\tau(x)$ (also refers to 'traction') according to the Coulomb friction law. The tangential sliding pressure distribution here is not the real state of the interface, but rather a limit state where the entire interface is in a sliding condition. The distribution function of traction is expressed as

$$\tau(x) = \mu p(N, x, a), \ |x| \leqslant a. \tag{5.5}$$

The IDF determines the critical value (or threshold) of the sliding force on the Jenkins element. This critical value, in turn, distinguishes between the stick and sliding regions in the contact area, thereby determining the distribution of the actual contact force.

Typically, the contact pressure distribution function is symmetrical. Therefore, to simplify the derivation of the model, the distribution of sliding stress is reconstructed, as shown in step 3 of figure 5.3. It is worth emphasizing that this reconstruction process does not alter the total sliding force. The reconstructed sliding tress distribution function is as follows,

$$\tau'(x) = 2\mu p(N, x, a), \ 0 \leqslant x \leqslant a. \tag{5.6}$$

In order to link the sliding stress distribution function with the framework of the Iwan model, the contact coordinates are normalized. This normalization enables the mapping of the sliding stress from spatial distribution to distribution on Jenkins elements. The sliding stress distributed in spatial coordinates has the following relationship with the critical sliding force distributed on Jenkins elements:

$$f^*(i) = a \cdot \tau'(x). \tag{5.7}$$

In statistical terms, $\varphi(f^*)\mathrm{d}f^*$ refers to the probability that the critical slip force falls within the range $[f^*, f^* + \mathrm{d}f^*]$, which is the probability of event $f^*$ occurring in the spatial domain:

$$\varphi(f^*)\mathrm{d}f^* = \frac{\mathrm{d}x}{a}. \tag{5.8}$$

From equation (5.8), the expression for the IDF can be derived. It is defined as the cotangent of the reconstructed tangential sliding stress distribution function in normalized coordinates, that is

$$\varphi(f^*) = \frac{\mathrm{d}x}{a \cdot \mathrm{d}f^*} = \left| \frac{\mathrm{d}\left(\frac{x}{a}\right)}{\mathrm{d}f^*} \right|. \tag{5.9}$$

The final step is to substitute the obtained IDF into equations (5.2) and (5.3) to derive the force–displacement expression.

Compared with the original Iwan model, the proposed generalized Iwan model does not contain parameters related to the assumed density function form. It includes only two model parameters: tangential contact stiffness and friction coefficient. These parameters can be extracted from experimental data. The slope of the initial stick stage in the hysteresis curve is defined as the tangential contact stiffness, and the ratio of the tangential force to the normal force in the gross slip regime is defined as the friction coefficient.

### 5.2.3 Effect of contact pressure distribution on friction

This section shows the effect of contact pressure distribution on interface friction behavior by comparing three typical distribution functions: uniform, linear, and Hertzian. The aim is to determine whether it is necessary to consider the contact pressure distribution when performing phenomenological modeling of partial slip friction.

In terms of the uniform distribution, all Jenkins elements have the same critical sliding force, which means that in this case, the Iwan model is equivalent to the gross slip model. The linear distribution can be considered a rough approximation of the Hertzian distribution. Figure 5.4 illustrates these three pressure distribution functions and corresponding IDFs. Table 5.1 lists the expression of friction force for different pressure distributions under monotonic loads.

A set of numerical simulations is conducted on the hysteresis loops of these three distributions under a constant normal load. The parameters for this numerical example include: tangential contact stiffness $k_t = 1 \times 10^7 \text{ N m}^{-1}$, friction coefficient $\mu = 0.5$ and a sinusoidal tangential motion $\delta = 10 \sin(2\pi t)\ \mu\text{m}$.

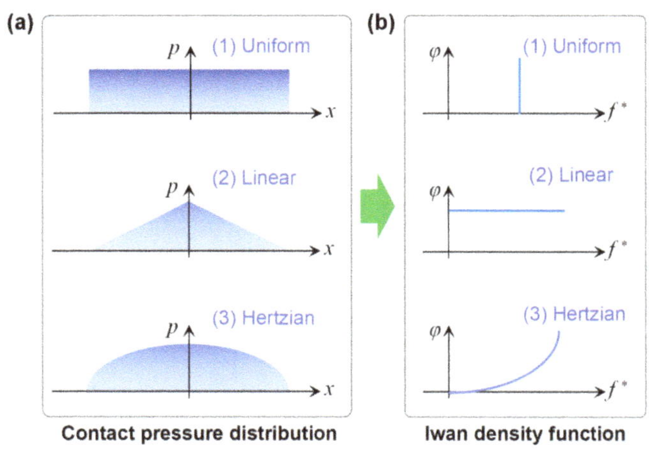


**Figure 5.4.** (a) Three typical contact pressure distribution functions (i.e. uniform, linear, and Hertzian distribution) and (b) the corresponding density functions of critical sliding force. (Reproduced with permission from [25]. Copyright 2019 Elsevier.)

Two cases of normal load are examined, 100 and 150 N. Figure 5.5 plots the friction force as a function of the tangential relative displacement for different pressure distribution functions. All three models reach the gross slip state for the case of 100 N. The uniform distribution model has a maximum enclosed area, which serves as a reference against that calculated by the others, as listed in table 5.2. The linear distribution model is the last to reach the gross slip state, with a minimum expected damping which is about 30% lower than that of the uniform distribution model. The Hertzian distribution model predicts a damping that is somewhere in between.

**Table 5.1.** The expression of friction force for different pressure distributions under monotonic loads. (Reproduced with permission from [25]. Copyright 2019 Elsevier.)

| Pressure distribution | Uniform | Linear | Hertzian |
|---|---|---|---|
| Friction force | $T(\delta) = k_t\delta$ | $T(\delta) = k_t\delta - \frac{(k_t\delta)^2}{4\mu N_0}$ | (5.22) |

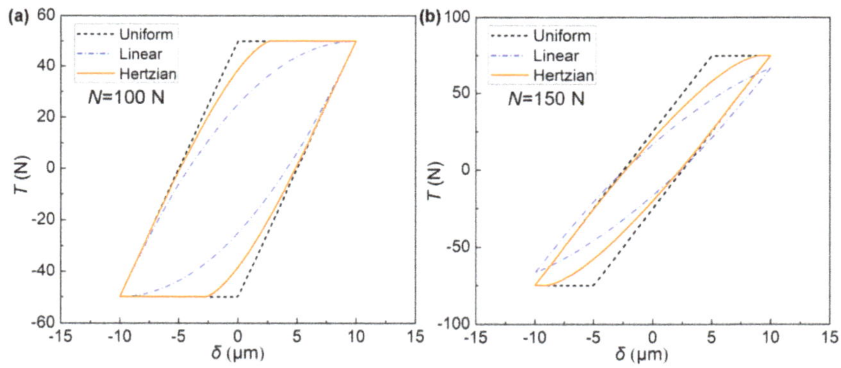


**Figure 5.5.** Comparison of hysteresis curves between different pressure distribution functions under two normal loads: (a) $N = 100$ N and (b) $N = 150$ N. (Reproduced with permission from [25]. Copyright 2019 Elsevier.)

**Table 5.2.** Comparison of the area enclosed by hysteresis loops between different contact pressure distribution functions. (Reproduced with permission from [25]. Copyright 2019 Elsevier.)

| Normal load (N) | Uniform (mJ) | Linear | Hertzian |
|---|---|---|---|
| 100 | 1 | 37% | 8% |
| 150 | 0.75 | 41.3% | 24% |

Similarly, for the case of 150 N, the friction damping can differ by as much as 40% due to variations in pressure distribution. Therefore, it can be concluded that the contact pressure distribution has a significant impact on the prediction results.

## 5.3 Analytical validation in a spherical contact

This section uses the generalized Iwan model to simulate the relationship between tangential frictional force and relative displacement when two elastic spheres are in contact. The prediction results of the model are compared with the Mindlin analytical solution [29, 32] to verify the model effectiveness.

### 5.3.1 Mindlin analytical solution

Mindlin [29, 32] studied the friction contact problem between two spheres under a constant normal load and a cyclic tangential force, as shown in figure 5.6. He considered the contact area to be a circle with radius $a$ and divided the contact area into a slip zone and stick zone. In the stick region, the tangential deformation is uniform, and interface sliding starts at the edge of the contact area, gradually extending towards the contact center as the tangential load increases.

Based on the theory of elasticity and Hertz contact theory, Mindlin deduced the relationship between tangential frictional force and displacement for spherical contact under a monotonically increasing tangential load, as follows:

$$\delta = \frac{3(2-\vartheta)\mu N}{16Ga}\left[1-\left(1-\frac{T}{\mu N}\right)^{2/3}\right], \tag{5.10}$$

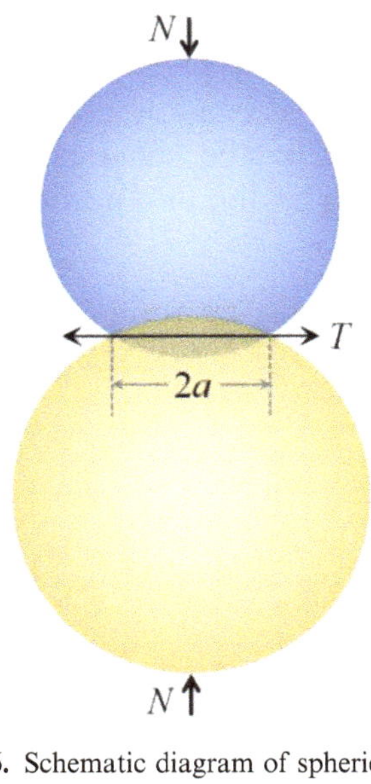


**Figure 5.6.** Schematic diagram of spherical contact.

where $\vartheta$ and $G$ are the Poisson's ratio and shear modulus of the material, respectively. The force–displacement relationship during the unloading and reloading phases under cyclic loads is also given as

$$\delta = \begin{cases} \dfrac{3(2-\vartheta)\mu N}{16Ga}\left[2\left(1-\dfrac{T_m-T}{2\mu N}\right)^{2/3}-\left(1-\dfrac{T_m}{\mu N}\right)^{2/3}-1\right], & \dot{T}<0 \\ -\dfrac{3(2-\vartheta)\mu N}{16Ga}\left[2\left(1-\dfrac{T_m+T}{2\mu N}\right)^{2/3}-\left(1-\dfrac{T_m}{\mu N}\right)^{2/3}-1\right], & \dot{T}>0 \end{cases}, \tag{5.11}$$

in which $T_m$ is the amplitude of friction force.

According to equation (5.11), the relationship between energy dissipation per cycle and the amplitude of friction force can be derived as

$$W = \frac{9(2-\vartheta)\mu^2N^2}{10Ga}\left\{\left[1-\left(1-\frac{T_m}{\mu N}\right)^{5/3}\right]-\frac{5T_m}{6\mu N}\left[1+\left(1-\frac{T_m}{\mu N}\right)^{2/3}\right]\right\}. \tag{5.12}$$

For a small $T_m/\mu N$, equation (5.12) is simplified to

$$W = \frac{(2-\vartheta)T_m^3}{36Ga\mu N}. \tag{5.13}$$

It can be seen from this formula that when the force amplitude is small, the energy dissipation per cycle in the Mindlin solution is related to the load amplitude by a cubic power. Note that this equation does not take into account the gross slip case and is only applicable to small tangential loads.

### 5.3.2 Modeling sphere-on-sphere contact

This section employs the generalized Iwan model to simulate the friction hysteresis behavior in spherical contact. For the spherical contact problem, the contact pressure distribution function can be obtained according to the Hertz contact theory [22], as shown in figure 5.7(a),

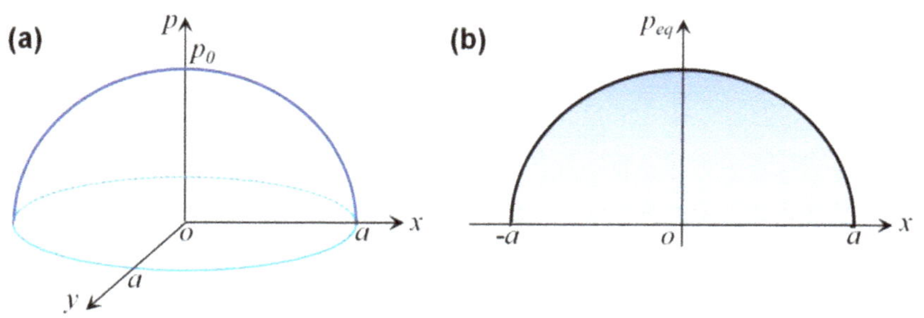


**Figure 5.7.** Contact pressure distribution of two elastic spheres: (a) on two-dimensional space and (b) on one-dimensional space. (Adapted with permission from [13]. Copyright 2020 Elsevier.)

$$p(r) = p_0\sqrt{1 - \left(\frac{r}{a}\right)^2}, \quad 0 \leqslant r \leqslant a, \tag{5.14}$$

where $r$ denotes the distance from any point in the contact area to the contact center, and $p_0$ is the contact pressure at contact center, $p_0 = 3N/(2\pi a^2)$.

The Iwan model is one-dimensional, whereas the pressure distribution of a spherical contact is in a two-dimensional spatial domain. Therefore, it is necessary to map the contact pressure distribution from the spatial coordinate system to the Jenkins element coordinate system.

To achieve this goal, the two-dimensional pressure distribution function of the spherical contact is equivalently converted to a one-dimensional distribution. The so-called 'equivalent' here means that the total contact pressure remains unchanged before and after the conversion. It can be implemented by integrating the contact pressure along the $y$-axis, as shown in figure 5.7(b). The equivalent pressure distribution function after conversion is

$$p_{\text{eq}}(x) = \int_{-\sqrt{a^2-x^2}}^{\sqrt{a^2-x^2}} p_0\sqrt{1 - \left(\frac{x}{a}\right)^2 - \left(\frac{y}{a}\right)^2}\,\mathrm{d}y = \frac{\pi p_0}{2a}(a^2 - x^2), \tag{5.15}$$

where $p_{\text{eq}}$ represents the line contact pressure, which is defined as the normal force per contact width. In this process, the two-dimensional circular contact area is compressed to one dimension, and the line contact pressure at the contact center is $\pi p_0 a/2$.

According to step 2 illustrated in section 5.2.2, substituting equation (5.15) into the Coulomb law yields the tangential sliding stress distribution function,

$$\tau(x) = \mu n_{\text{eq}}(x) = \frac{\pi\mu p_0}{2a}(a^2 - x^2), \quad -a \leqslant x \leqslant a. \tag{5.16}$$

Since the sliding stress distribution is symmetrical, the contact segments with the same sliding stress are sorted for reconstruction. The reconstructed sliding pressure distribution function is

$$\tau'(x) = \frac{\pi\mu p_0}{a}(a^2 - x^2), \ 0 \leqslant x \leqslant a. \tag{5.17}$$

Then the coordinates in the contact region are normalized, and the normalized sliding stress distribution function is written as

$$\tau'\left(\frac{x}{a}\right) = \pi\mu p_0 a\left[1 - \left(\frac{x}{a}\right)^2\right], \quad 0 \leqslant \frac{x}{a} \leqslant 1. \tag{5.18}$$

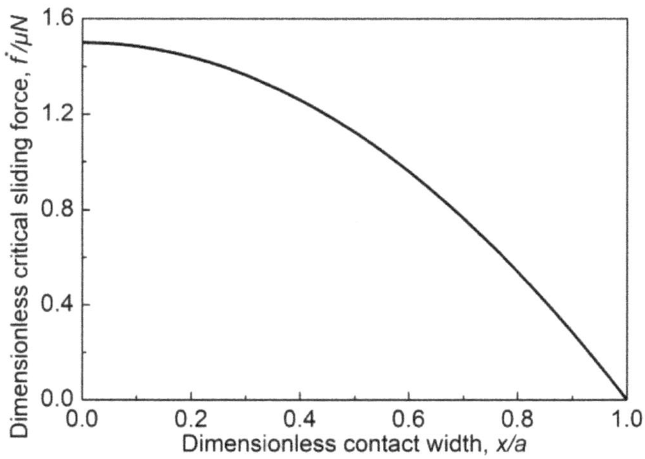


**Figure 5.8.** Dimensionless critical sliding force $f^*/\mu N$ versus dimensionless contact width $x/a$. (Adapted with permission from [13]. Copyright 2020 Elsevier.)

The critical sliding force in the normalized contact coordinate is then

$$f^*\left(\frac{x}{a}\right) = a \cdot \tau'\left(\frac{x}{a}\right) = \pi\mu p_0 a^2\left[1 - \left(\frac{x}{a}\right)^2\right]. \tag{5.19}$$

Figure 5.8 plots the dimensionless critical sliding force $f^*/\mu N$ as a function of dimensionless contact width $x/a$. Substituting $p_0 = 3N/(2\pi a^2)$ into equation (5.19) yields the normalized contact coordinate,

$$\frac{x}{a} = \sqrt{1 - \frac{2f^*}{3\mu N}},\quad 0 \leqslant \frac{x}{a} \leqslant 1. \tag{5.20}$$

Finally, according to the definition of the IDF, the expression of the IDF can be obtained by differentiating equation (5.20),

$$\varphi(f^*) = \left|\frac{\mathrm{d}\left(\frac{x}{a}\right)}{\mathrm{d}f^*}\right| = \frac{1}{\sqrt{3\mu N(3\mu N - 2f^*)}}. \tag{5.21}$$

Figure 5.9 plots the dimensionless density function $\varphi(f^*)\mu N$ as a function of the dimensionless critical sliding force $f^*/\mu N$. Substituting equation (5.21) into equation (5.2) yields the friction force under small load conditions,

$$T = \mu N - \frac{(3\mu N - 2k_t\delta)^{\frac{3}{2}}}{3\sqrt{3\mu N}}. \tag{5.22}$$

For cyclic loading, the friction force can be obtained based on the Masing hypothesis.

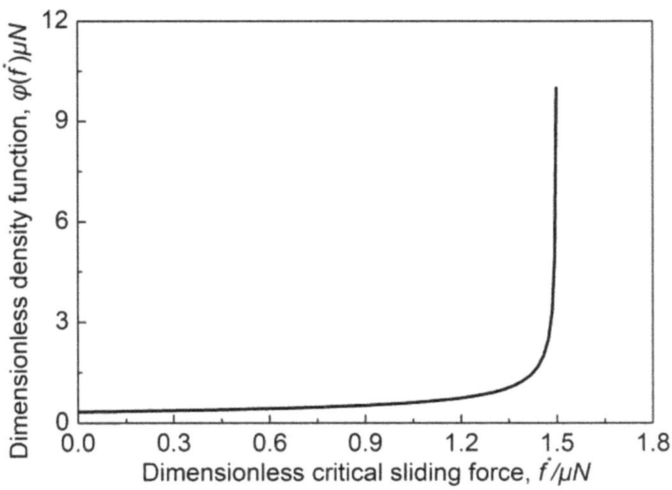


**Figure 5.9.** Dimensionless density function $\varphi(f^*)\mu N$ versus dimensionless critical sliding force $f^*/\mu N$. (Adapted with permission from [13]. Copyright 2020 Elsevier.)

**Table 5.3.** Model parameters and loading conditions. (Reproduced with permission from [13]. Copyright 2020 Elsevier.)

| Shear modulus | Poisson's ratio | Friction coefficient | Contact radius | Normal load |
|---|---|---|---|---|
| 27 GPa | 0.33 | 0.5 | 2 mm | 500 N |

### 5.3.3 Comparison with the analytical solution

The following example compares the predictions of the generalized Iwan model with the Mindlin analytical solution to validate the effectiveness of the model. This example calculates the friction response for two identical elastic spheres under a constant normal load and a periodic tangential load.

The material parameters, friction coefficient, and normal load of the sphere are listed in table 5.3. The tangential load is $T = T_m \sin(2\pi t)$. The tangential contact stiffness to be used in the model is the slope of the force–displacement curve when the contacting body is at the onset of relative motion. For consistency, this stiffness can be obtained according to the limit of the derivative of the friction force with respect to the relative displacement,

$$k_t = \lim_{\delta \to 0} \frac{\partial T}{\partial \delta} = \lim_{\delta \to 0} \frac{8Ga}{2 - \vartheta} \left[ 1 - \frac{16Ga\delta}{3(2 - \vartheta)\mu N} \right]^{\frac{1}{2}} = \frac{8Ga}{2 - \vartheta}. \tag{5.23}$$

The displacement response calculated by the Mindlin analytical solution (equation (5.11)) is used as the input of the generalized Iwan model. Two cases are

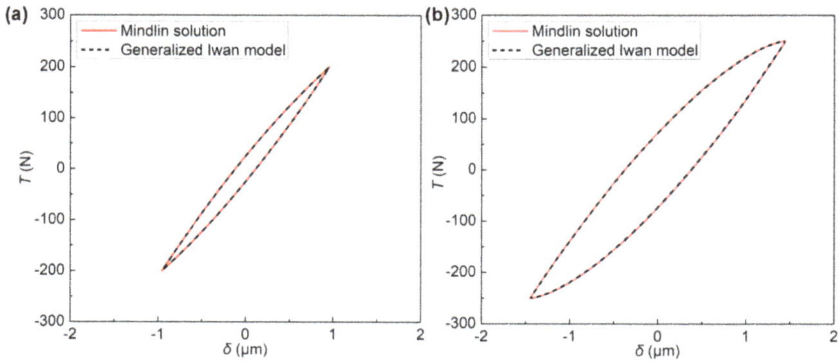


**Figure 5.10.** Comparison of hysteresis loops simulated by the generalized Iwan model and the Mindlin solution: (a) $T_m = 200$ N and (b) $T_m = 250$ N. (Adapted with permission from [13]. Copyright 2020 Elsevier.)

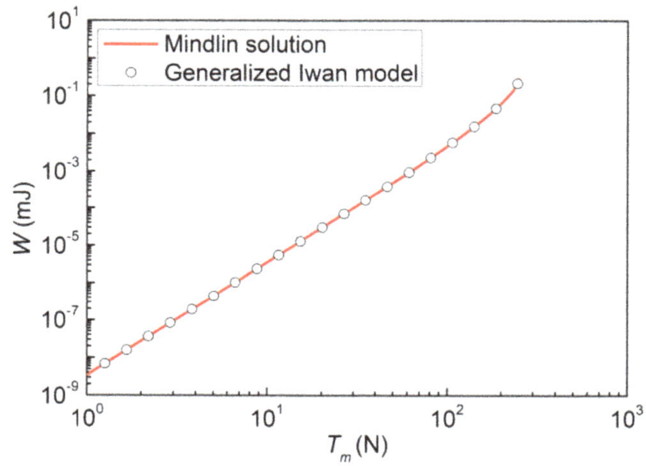


**Figure 5.11.** Comparison of energy dissipation predicted by the generalized Iwan model and the Mindlin solution. (Adapted with permission from [13]. Copyright 2020 Elsevier.)

examined: a small relative displacement for which the sliding condition is not reached (maximum friction force $T_m = 200 \text{ N} < \mu N$) and a large relative displacement up to gross slip conditions (maximum friction force $T_m = 250 \text{ N} = \mu N$).

Figure 5.10 compares the predicted hysteresis lops with the analytical solutions. These two sets of curves completely overlapped, as can be easily verified by substituting equation (5.23) into equation (5.22).

Figure 5.11 compares the dissipated energy per cycle and confirms the equivalence of the two methods. The generalized Iwan model can also be used to simulate friction behavior of other contact geometries. In the next section, a lap joint is modeled using the generalized Iwan model.

## 5.4 Experimental validation in a plane contact

Plane contact scenarios are widely present in assembled structures, such as lap joints connected by bolts, rivets, etc. The frictional contact behavior at these joints significantly affects the dynamic behavior of the overall assembled structure. Therefore, developing reliable friction models for plane contact is essential for accurately predicting the dynamic response of these structures. This section uses the generalized Iwan model to simulate the contact of a lap joint plate and validates it by comparison with experimental results.

Figure 5.12 shows a schematic diagram of the plane lap plate model. The normal load applied at the center of the lap area causes the two plates to contact with each other and reciprocate under the action of the tangential load. The normal load along the width direction is assumed to be uniformly distributed. The length of the contact area is $a$ and the cross section of the plate is a rectangle with width $w$ and thickness $h$.

### 5.4.1 Modeling friction hysteresis in lap joints

It is challenging to obtain an accurate distribution function of contact pressure for plane contact problems. Allarà [33] provided an analytical solution of the contact pressure distribution for plane contact bodies with rounded corners. However, in actual manufacturing processes, the contact surface exhibits waviness and micro-asperities, which can significantly affect the contact pressure distribution and contact area. Moreover, cyclic loading modifies the contact due to changes in surface topography and redistribution of contact pressure. These phenomena, more pronounced in plane contact problems than in spherical contact, could limit the application of the generalized Iwan model.

To overcome this limitation, a quadratic function with a variable parameter $\gamma$ is used to assume the distribution function of contact pressure. This variable parameter is denoted ‘proportional coefficient’, which is defined as the ratio of the peak pressure $p_0$ at the contact center to the mean pressure $p_m = N/aw$,

$$\gamma = \frac{awp_0}{N}. \tag{5.24}$$

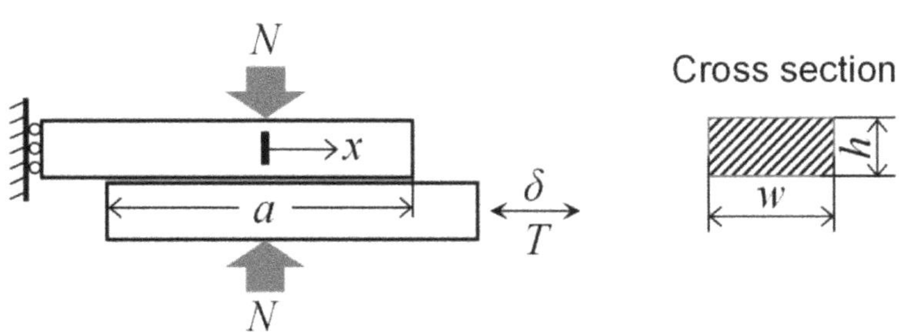


**Figure 5.12.** Lap joint plate model and cross section of the plate. (Adapted with permission from [13]. Copyright 2020 Elsevier.)

This quadratic function can represent three typical pressure distribution forms, i.e. 'concave', 'convex', and 'uniform' distributions. These distribution forms are widely used to approximate real situations [34]. An optimal choice of the proportional coefficient can be inferred from measured results.

Since the normal load is concentrated at the center of the plate, the extreme value of the contact pressure must be located at the center of the contact region. Constrained by this condition, the expression of the assumed distribution function of contact pressure is unique,

$$p(x) = \frac{N}{aw}\left[\frac{12(1-\gamma)x^2}{a^2} + \gamma\right], \quad x \leqslant \left|\frac{a}{2}\right|, \tag{5.25}$$

where $\gamma$ is the proportional coefficient and ranges from 0 to 1.5. For the case $\gamma < 1$, equation (5.25) represents a concave distribution, while for the case $\gamma > 1$, it is a convex distribution function. When $\gamma = 1$, the contact pressure is evenly distributed.

Figure 5.13 shows dimensionless contact pressure $p(x)aw/N$ distributed on normalized spatial coordinate *x/a* for different proportional coefficients. For a practical lap joint structure, the corresponding pressure distribution function can be approximated by selecting an appropriate proportional coefficient.

According to the generalized Iwan model, the tangential sliding stress distribution can be written as

$$\tau(x) = \mu p(x) = \frac{\mu N}{aw}\left[\frac{12(1-\gamma)x^2}{a^2} + \gamma\right], \quad x \leqslant \left|\frac{a}{2}\right|. \tag{5.26}$$

Then, the contact coordinates are normalized, and the symmetry of the distribution function is considered. The tangential sliding pressure distribution is reconstructed as

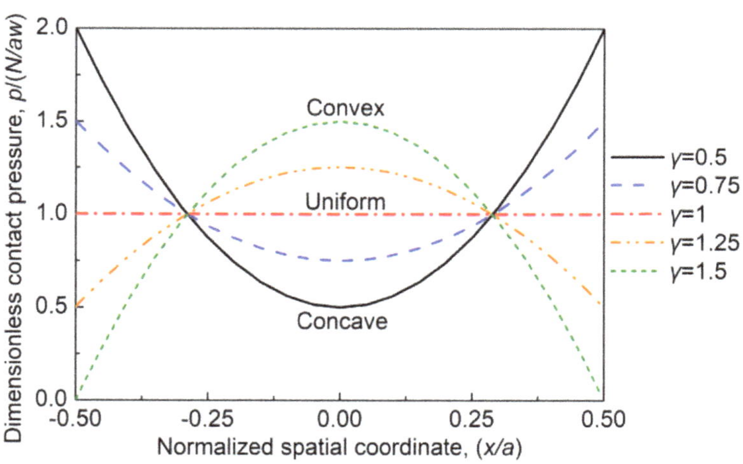


**Figure 5.13.** Dimensionless contact pressure *paw/N* versus normalized contact coordinate *x/a* for different proportional coefficients. (Adapted with permission from [13]. Copyright 2020 Elsevier.)

$$\tau'\left(\frac{2x}{a}\right) = \frac{2\mu N}{aw}\left[3(1-\gamma)\left(\frac{2x}{a}\right)^2 + \gamma\right], \quad 0 \leqslant \frac{2x}{a} \leqslant 1. \tag{5.27}$$

The critical sliding force in the normalized contact coordinate is

$$f^*\left(\frac{2x}{a}\right) = \frac{aw}{2}\cdot\tau'\left(\frac{2x}{a}\right) = \mu N\left[3(1-\gamma)\left(\frac{2x}{a}\right)^2 + \gamma\right]. \tag{5.28}$$

Correspondingly, the IDF can be derived as

$$\varphi(f^*) = \left|\frac{\mathrm{d}\left(\frac{2x}{a}\right)}{\mathrm{d}f^*}\right| = \frac{1}{2\mu N\sqrt{3(1-\gamma)\left(\frac{f^*}{\mu N} - \gamma\right)}}. \tag{5.29}$$

Figure 5.14 plots the dimensionless IDF $2\mu N\varphi(f^*)$ as a function of dimensionless critical sliding force $f^*/\mu N$ for different proportional coefficients. For the concave distribution, the IDF approaches infinity when the dimensionless critical sliding force approaches its minimum, $\gamma\mu N$. Whereas for the convex distribution, the IDF goes to infinity when the dimensionless critical sliding force approaches its maximum.

Finally, substitute the IDF into equation (5.2) to obtain the integral expression of friction force,

$$T(\delta) = \int_{\mathrm{lb}}^{k_t\delta} f^*\varphi(f^*)\mathrm{d}f^* + k_t\delta\int_{k_t\delta}^{\mathrm{ub}} \varphi(f^*)\mathrm{d}f^*, \tag{5.30}$$

where lb and ub denote the lower and upper bounds of the integral, respectively. These values are different between the concave and convex distributions, as listed in table 5.4.

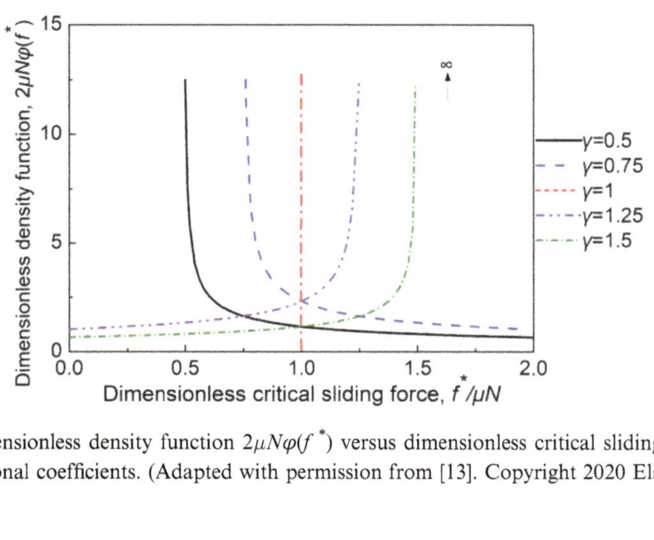


**Figure 5.14.** Dimensionless density function $2\mu N\varphi(f^*)$ versus dimensionless critical sliding force $f^*/\mu N$ for different proportional coefficients. (Adapted with permission from [13]. Copyright 2020 Elsevier.)

**Table 5.4.** Upper and lower bounds of the friction formula for the concave and convex distributions. (Adapted with permission from [13]. Copyright 2020 Elsevier.)

| Variables | Concave distribution ($\gamma < 1$) | Convex distribution ($\gamma > 1$) |
|---|---|---|
| Upper bound | $\mathrm{ub} = (3 - 2\gamma)\mu N$ | $\mathrm{ub} = \gamma\mu N$ |
| Lower bound | $\mathrm{lb} = \gamma\mu N$ | $\mathrm{lb} = (3 - 2\gamma)\mu N$ |

Substituting the upper and lower bounds listed in table 5.4 into equation (5.30) yields the explicit expression of friction force. For the case of a concave distribution,

$$T(\delta) = \begin{cases} k_t\delta, & \delta \leqslant \dfrac{(3 - 2\gamma)\mu N}{k_t} \\ \mu N - \dfrac{2(\gamma\mu N - k_t\delta)}{3}\sqrt{\dfrac{k_t\delta - \gamma\mu N}{3(1 - \gamma)\mu N}}, & \dfrac{(3 - 2\gamma)\mu N}{k_t} < \delta \leqslant \dfrac{\gamma\mu N}{k_t} \\ \mu N, & \delta > \dfrac{\gamma\mu N}{k_t} \end{cases}. \quad (5.31)$$

For the case of a convex distribution,

$$T(\delta) = \begin{cases} k_t\delta, & \delta \leqslant \dfrac{\gamma\mu N}{k_t} \\ k_t\delta + \dfrac{2(\gamma\mu N - k_t\delta)}{3}\sqrt{\dfrac{k_t\delta - \gamma\mu N}{3(1 - \gamma)\mu N}}, & \dfrac{\gamma\mu N}{k_t} < \delta \leqslant \dfrac{(3 - 2\gamma)\mu N}{k_t} \\ \mu N, & \delta > \dfrac{(3 - 2\gamma)\mu N}{k_t} \end{cases}. \quad (5.32)$$

### 5.4.2 Model validation

The experimental results from the second-generation fretting test device at the AERMEC laboratory of the Polytechnic University of Turin were used for comparative validation. This device measures the relative displacement between the fixed and mobile specimens using two laser vibrometers, and the tangential friction force using a load cell. It has good accuracy and repeatability. The specific experimental method are detailed in [35], and will not be further elaborated here. Figure 5.15 shows the fixed and mobile specimens and the contact area, where the nominal contact area is two rectangles with an area of 5 mm$^2$.

It is worth noting that in some experimental results, a 'residual stiffness' phenomenon is observed during the gross slip stage. Therefore, the generalized Iwan model needs to be further extended to incorporate the simulation of residual stiffness. The method used here involves paralleling a linear spring with stiffness $\alpha k_t$ in the Iwan model, as shown in figure 5.16. In this case, the integral expression for the friction force becomes

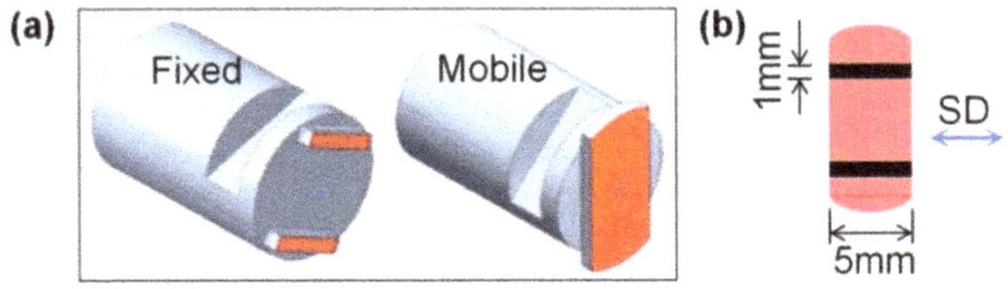


**Figure 5.15.** (a) Fixed and mobile specimens, with contact surfaces marked in red, and (b) contact area in black and sliding direction (SD) shown. (Adapted with permission from [13]. Copyright 2020 Elsevier.)

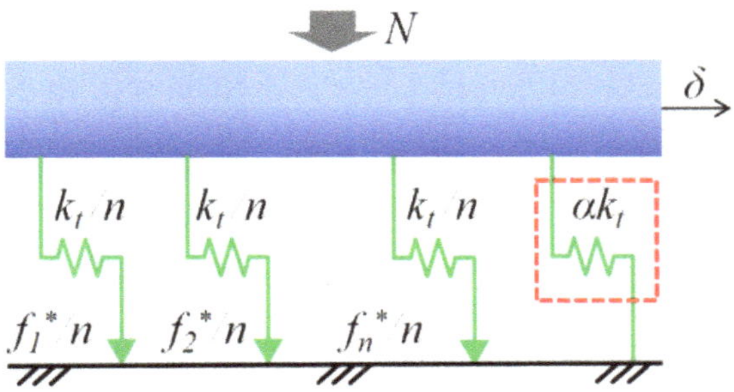


**Figure 5.16.** Generalized Iwan model considering residual contact stiffness.

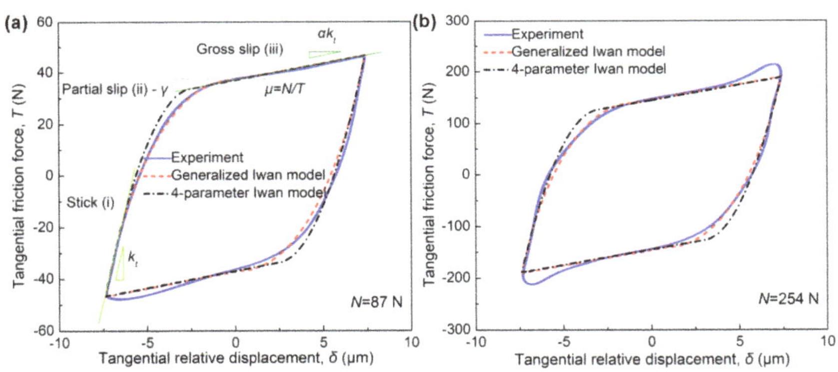


**Figure 5.17.** Comparison of experimental and simulated hysteresis curves under different normal loads: (a) $N = 87$ N and (b) $N = 254$ N. (Adapted with permission from [13]. Copyright 2020 Elsevier.)

$$T(\delta) = \int_0^{k_t\delta} f^* \varphi(f^*) d f^* + k_t\delta \int_{k_t\delta}^{\infty} \varphi(f^*) \mathrm{d}f^* + \alpha k_t \delta. \tag{5.33}$$

The model parameters, including tangential contact stiffness, friction coefficient, proportional coefficient, and residual stiffness coefficient, are extracted from measured hysteresis curves. Tests are conducted on two different samples under different normal loads, measuring the hysteresis curves at the contact surface. Figure 5.17 presents the experimental results (the blue solid lines) for the two conditions.

**Table 5.5.** Applied normal load and estimated model parameters. (Adapted with permission from [13]. Copyright 2020 Elsevier.)

| Normal load (N) | Tangential stiffness (N $\mu$m$^{-1}$) | Residual stiffness coefficient | Friction coefficient | Proportional coefficient |
|---|---|---|---|---|
| 87 | 31 | 0.043 | 0.48 | 0.2 |
| 254 | 134 | 0.045 | 0.65 | 0.2 |

To illustrate the extraction of model parameters, the case of $N = 87$ N is taken as an example. The measured hysteresis curve is divided into three motion stages: (i) stick, (ii) partial slip, and (iii) gross slip, as shown in figure 5.17(a). The slope of the curve in the stick stage is considered the tangential contact stiffness. The ratio of the interval between the curves in the forward and reverse gross slip stages to twice the normal load is defined as the friction coefficient. The ratio of the slope of the curve in the gross slip stage to the tangential contact stiffness is defined as the residual stiffness coefficient. Additionally, the proportional coefficient is extracted by comparing the overlap between simulated and experimental results in the partial stage. Table 5.5 lists the extracted model parameters for the two conditions.

Table 5.5 shows that the proportional coefficient is the same for these two cases, with a value of 0.2. Figure 5.17 compares the experimental hysteresis curves with the simulation results by the generalized Iwan model and the four-parameter Iwan model [26]. The parameters of the four-parameter Iwan model include tangential contact stiffness, critical sliding force, power exponent characterizing the energy dissipation, and residual stiffness coefficient. The power exponent is selected by fitting to minimize the error between the measured hysteresis curve and the simulation results. The optimal power exponent is 0.

The results show that the simulated hysteresis curves by the generalized Iwan model matches the measured curves better than that by the four-parameter Iwan model, especially in the partial slip regime. Figure 5.17(a) indicates a slight difference between the experimental and simulated results after the motion direction is reversed. This difference is due to different contact stiffness during the unloading and reloading stages, which sometimes occur in the test. Figure 5.17(b) shows that there are two uncommon bulges in the measured curves. The first bulge appears during the transition from the stick state to the partial slip state, possibly caused by local interactions between asperities on the contact surface. The second one occurs at the end of the gross slip stage, which is observed in many tests [36–38]. However, the cause has not yet been definitively explained. Currently, there are two main possible reasons: 'velocity effect' (interface velocity gradually reduces to zero at the end of the gross slip stage) and interactions at contact edges. This phenomenon is not yet considered in the Iwan model.

Due to the waviness and rough asperities on the contact surface caused by machining, it is difficult to analytically obtain a reliable contact pressure distribution. Even so, the results show that the generalized Iwan model based on a quadratic distribution with a varying proportional coefficient can yield satisfactory results.

### 5.4.3 Model parametric analysis

The model in section 5.4.1 introduces a proportional coefficient (varies in the range [0, 1.5]), which significantly affects the shape of hysteresis curves. This section analyzes the sensitivity of the hysteresis behavior and the energy dissipation per cycle to the proportional coefficient. Without loss of generality, the tangential relative displacement and friction force are normalized as $\delta_{\text{norm}} = k_t\delta/\mu N$, and $T_{\text{norm}} = T/\mu N$, respectively, and the corresponding force–displacement is written as

$$T_{\text{norm}} = \begin{cases} \delta_{\text{norm}} + \dfrac{2(\gamma - \delta_{\text{norm}})}{3}\sqrt{\dfrac{\delta_{\text{norm}} - \gamma}{3(1-\gamma)}}, & 0 \leqslant \gamma \leqslant 1 \\ 1 - \dfrac{2(\gamma - \delta_{\text{norm}})}{3}\sqrt{\dfrac{\delta_{\text{norm}} - \gamma}{3(1-\gamma)}}, & 1 < \gamma \leqslant 1.5 \end{cases}. \tag{5.34}$$

Figure 5.18 plots the normalized hysteresis curves for different proportional coefficients. The applied displacement is a sinusoidal wave with a normalized amplitude of 2 and a frequency of 1 Hz. The results show that the shape of the hysteresis curve is particularly sensitive to the proportional coefficients, especially in the gross slip stage. When the proportional coefficient is less than 1, the partial slip effect gradually weakens as the proportional efficient increases. In contrast, when the proportional efficient is greater than 1, the evolution trend is the opposite.

Figure 5.19 plots the normalized energy dissipation per cycle (energy dissipation divided by its maximum value) as a function of the proportional coefficient for different displacement amplitudes. The normalized energy dissipation is not monotonic and shows dependence on amplitudes of displacement. For a relatively

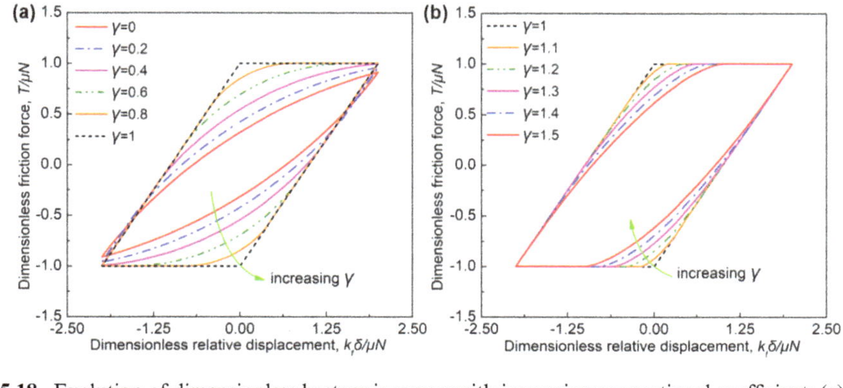


**Figure 5.18.** Evolution of dimensionless hysteresis curves with increasing proportional coefficient: (a) $\gamma$ in the range [0, 1] and (b) $\gamma$ in the range [1, 1.5]. (Adapted with permission from [13]. Copyright 2020 Elsevier.)

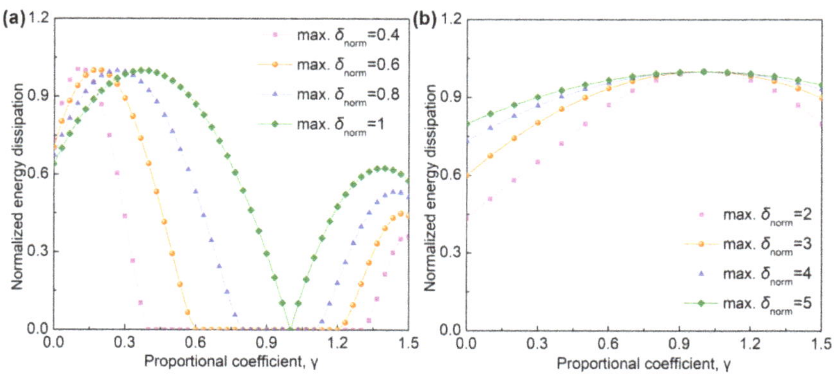


**Figure 5.19.** Normalized energy dissipation (enclosed area divided by its maximum value) versus proportional coefficient under different displacement amplitudes. (Adapted with permission from [13]. Copyright 2020 Elsevier.)

small displacement amplitude, the energy dissipation is zero within a certain range of proportional coefficients. In this case, the stick regime dominates the motion of contact surfaces, and the smaller the displacement amplitude, the wider the range of proportional coefficients corresponding to zero energy dissipation, while for a relatively larger displacement amplitude, the energy dissipation reaches a peak at $\gamma = 1$. As the displacement amplitude increases, the impact of the proportional coefficient on energy dissipation gradually decreases because the contact transitions to a gross slip state.

## 5.5 Summary

The Iwan model is a representative phenomenological model for interface friction and is widely used in practical engineering. The core of the Iwan model is the IDF of the critical sliding force, which determines the prediction accuracy of the model. The traditional Iwan-type models assumes the IDF as a uniform distribution or a truncated power-law distribution. However, these assumptions do not make any physical sense. This chapter introduces a generalized Iwan model that relates the contact pressure distribution to the IDF. This method gives the IDF an explicit physical meaning. The main results and conclusions are summarized as follows:

1. The mathematical relationship between the contact pressure distribution and the IDF is constructed, and the expression of the IDF is derived in detail. Then the IDF is incorporated into the Iwan model framework to obtain the force–displacement expressions for the monotonic and cyclic loads.
2. The generalized Iwan model is employed to simulate the friction hysteresis of spherical contacts. Based on Hertz contact theory, the IDF and corresponding force–displacement relation are derived. The comparison with the Mindlin analytical solution shows that the two models produce identical

hysteresis curve and energy dissipation, thus validating the effectiveness of the generalized Iwan model.

3. Finally, the generalized Iwan model is applied to a planar contact of lap joints. Given the difficulty in obtaining the pressure distribution function for planar contact, a quadratic function with a variable proportional coefficient is assumed. The proportional coefficient significantly affects the shape of the hysteresis curve, especially in the partial slip stage. In terms of experimental verification, the proportional coefficient remains the same or shows a non-appreciable variation for different normal loads. The simulated results of the generalized Iwan model are in good agreement with the experimental counterparts and are closer to the experiments than those of the four-parameter Iwan model. This indicates that the generalized Iwan model is effective and more accurate.

## Bibliography


[1] Iwan W D 1966 A distributed-element model for hysteresis and its steady-state dynamic response *ASME J. Appl. Mech.* **33** 893–900

[2] Iwan W D 1967 On a class of models for the yielding behavior of continuous and composite systems *ASME J. Appl. Mech.* **34** 612–7

[3] Gaul L and Lenz J 1997 Nonlinear dynamics of structures assembled by bolted joints *Acta Mech.* **125** 169–81

[4] Abad J, Medel F J and Franco J M 2014 Determination of Valanis model parameters in a bolted lap joint: experimental and numerical analyses of frictional dissipation *Int. J. Mech. Sci.* **89** 289–98

[5] Jalali H, Jamia N, Friswell M I, Khodaparast H H and Taghipour J 2022 A generalization of the Valanis model for friction modelling *Mech. Syst. Sig. Process.* **179** 109339

[6] Ikhouane F, Mañosa V and Rodellar J 2007 Dynamic properties of the hysteretic Bouc–Wen model *Syst. Control Lett.* **56** 197–205

[7] Ismail M, Ikhouane F and Rodellar J 2009 The hysteresis Bouc–Wen model, a survey *Arch. Comput. Meth. Eng.* **16** 161–88

[8] Ikhouane F and Rodellar J 2007 *Systems with Hysteresis: Analysis, Identification and Control Using the Bouc–Wen Model* (New York: Wiley)

[9] Song J and Der Kiureghian A 2006 Generalized Bouc–Wen model for highly asymmetric hysteresis *J. Eng. Mech.* **132** 610–8

[10] Dahl P R 1976 Solid friction damping of mechanical vibrations *AIAA J.* **14** 1675–82

[11] Johanastrom K and Canudas-De-Wit C 2008 Revisiting the LuGre friction model *IEEE Control Syst. Mag.* **28** 101–14

[12] Piatkowski T 2014 Dahl and LuGre dynamic friction models—the analysis of selected properties *Mech. Mach. Theory* **73** 91–100

[13] Li D, Botto D, Xu C and Goal M 2020 A new approach for the determination of the Iwan density function in modeling friction contact *Int. J. Mech. Sci.* **180** 105671

[14] Argatov I I and Butcher E A 2011 On the Iwan models for lap-type bolted joints *Int. J. Non Linear Mech.* **46** 347–56

[15] Zhan W and Huang P 2018 Physics-based modeling for lap-type joints based on the Iwan model *J. Tribol.* **140** 051401
[16] Segalman D J and Starr M J 2008 Inversion of Masing models via continuous Iwan systems *Int. J. Non Linear Mech.* **43** 74–80
[17] Song Y, Hartwigsen C J, McFarland D M, Vakakis A F and Bergman L A 2004 Simulation of dynamics of beam structures with bolted joints using adjusted Iwan beam elements *J. Sound Vib.* **273** 249–76
[18] Wang D, Xu C, Fan X and Wan Q 2018 Reduced-order modeling approach for frictional stick–slip behaviors of joint interface *Mech. Syst. Sig. Process.* **103** 131–8
[19] Li Y and Hao Z 2016 A six-parameter Iwan model and its application *Mech. Syst. Sig. Process.* **68** 354–65
[20] Li Y, Hao Z, Feng J and Zhang D 2017 Investigation into discretization methods of the six-parameter Iwan model *Mech. Syst. Sig. Process.* **85** 98–110
[21] Brake M R W 2017 A reduced Iwan model that includes pinning for bolted joint mechanics *Nonlinear Dyn.* **87** 1335–49
[22] Johnson K L 1987 *Contact Mechanics* (Cambridge: Cambridge University Press)
[23] Rajaei M and Ahmadian H 2014 Development of generalized Iwan model to simulate frictional contacts with variable normal loads *Appl. Math. Modell.* **38** 4006–18
[24] Li D, Xu C, Liu T, Gola M M and Wen L 2019 A modified IWAN model for micro-slip in the context of dampers for turbine blade dynamics *Mech. Syst. Sig. Process.* **121** 14–30
[25] Li D, Botto D, Xu C, Liu T and Gola M M 2019 A micro-slip friction modeling approach and its application in underplatform damper kinematics *Int. J. Mech. Sci.* **161** 105029
[26] Segalman D J 2005 A four-parameter Iwan model for lap-type joints. ASME *J. Appl. Mech.* **72** 752–60
[27] Ranjan P and Pandey A K 2021 Modeling of pinning phenomenon in Iwan model for bolted joint *Tribol. Int.* **161** 107071
[28] Shen M M, Yang X D, Gao C, Yang J H, Shi R and Guo P F 2024 Modeling and analyzing the influence of slip velocity on joint surface *Int. J. Non Linear Mech.* **165** 104798
[29] Mindlin R D, Mason W P, Osmer T F and Deresiewicz H 1952 Effects of an oscillating tangential force on the contact surfaces of elastic spheres *Proc. of the 1st US National Congress of Applied Mechanics* vol 1951 pp 203–8
[30] Eriten M, Polycarpou A A and Bergman L A 2011 Effects of surface roughness and lubrication on the early stages of fretting of mechanical lap joints *Wear* **271** 2928–39
[31] Jenkins G M 1962 Analysis of the stress–strain relationships in reactor grade graphite *Br. J. Appl. Phys.* **13** 30
[32] Mindlin R D and Deresiewicz H 1953 Elastic spheres in contact under varying oblique forces *ASME J. Appl. Mech.* **20** 327–44
[33] Allarà M 2009 A model for the characterization of friction contacts in turbine blades *J. Sound Vib.* **320** 527–44
[34] Cigeroglu E, Lu W and Menq C H 2006 One-dimensional dynamic microslip friction model *J. Sound Vib.* **292** 881–98
[35] Lavella M, Botto D and Gola M M 2011 Test rig for wear and contact parameters extraction for flat-on-flat contact surfaces *Int. Joint Tribology Conf.* vol 54747 pp 307–9

[36] Lavella M and Botto D 2011 Fretting wear characterization by point contact of nickel superalloy interfaces *Wear* **271** 1543–51

[37] Mulvihill D M, Kartal M E, Olver A V, Nowell D and Hills D A 2011 Investigation of non-Coulomb friction behaviour in reciprocating sliding *Wear* **271** 802–16

[38] Kartal M E, Mulvihill D M, Nowell D and Hills D A 2011 Measurements of pressure and area dependent tangential contact stiffness between rough surfaces using digital image correlation *Tribol. Int.* **44** 1188–98

# Chapter 6

# Modeling friction hysteresis at bolted joint interfaces

The friction between connected surfaces is one of the main sources of nonlinearity of the mechanical behaviors in bolted joint structures, and the energy dissipation induced by friction can account for up to 90% of the overall structural damping. This chapter uses the generalized Iwan model to characterize the nonlinear friction hysteresis behavior of bolted joint interfaces and applies it to predict the response of the bolted connection system under quasi-static and dynamic excitations, with cross-comparison and validation against measured results.

## 6.1 Contact pressure at bolted joint interfaces

This section conducts mechanical modeling of the friction hysteresis at bolted joint interfaces based on the framework of the generalized Iwan model given in chapter 5. Unlike spherical contact and cylindrical-plane contact, it is extremely challenging to analytically obtain the contact pressure distribution function of bolted connection surfaces. Here, the contact pressure distribution is thus calculated through a finite element model of a typical bolted connection and then fitted to obtain the continuous distribution function.

Two identical steel plates with dimensions $50 \times 40 \times 12$ mm$^3$ pre-tightened by an M8 bolt are modeled using the finite element (FE) software ABAQUS. The nominal contact area is 1286 mm$^2$. The material of the steel plates and bolt is assumed to be linear elastic. The material properties are set as follows: Young's modulus 210 GPa, Poisson's ratio 0.3. Each steel plate is meshed into 7840 eight-node reduced integration elements (C3D8R, element properties in ABAQUS). The contact interaction at connection surfaces is defined using a tangential 'Penalty' friction formulation and a normal 'Hard' contact algorithm. For more details, see [1].

Figure 6.1(a) shows the FE model of the bolted connection. The FE simulations were conducted for different preloads, 15, 18, and 20 kN. Figures 6.1(b)–(d) plot the

doi:10.1088/978-0-7503-6214-6ch6  

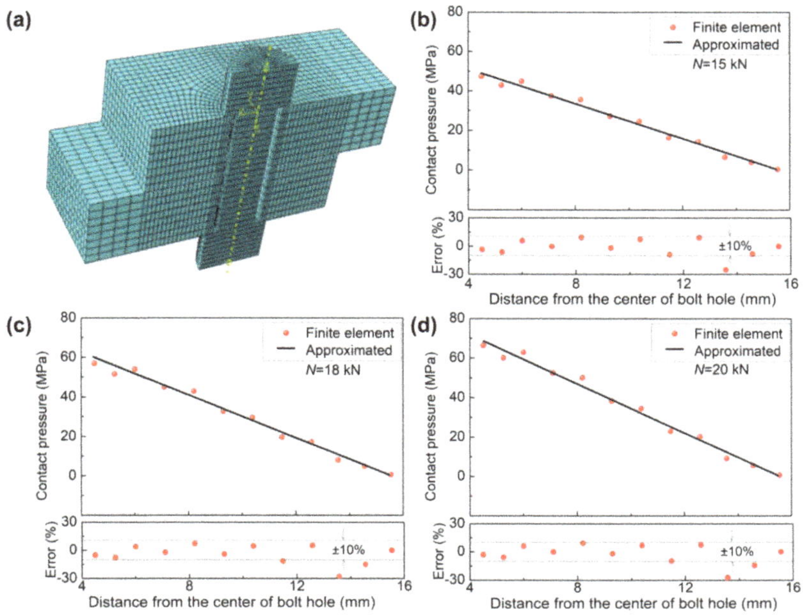


**Figure 6.1.** (a) Finite element contact model of an M8 bolt connection, and contact pressure distribution and relative error between the simulated and approximated results: (b) $N = 15$ kN, (c) $N = 18$ kN, and (d) $N = 20$ kN. (Adapted with permission from [1]. Copyright 2020 Springer Nature.)

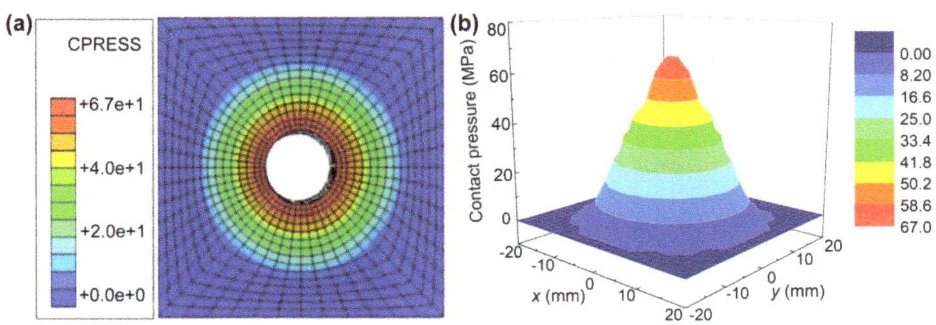


**Figure 6.2.** Contact stress distribution at bolted joint interfaces. (Adapted with permission from [1]. Copyright 2020 Springer Nature.)

contact pressure along the contact radius. It shows the simulated contact pressure decreases gradually from a maximum value on the edge of the bolt hole to zero on the boundary.

The single-bolt connection has an annulus contact area surrounding the through hole, as shown in figure 6.2(a) where the case of 20 kN preload is set as an example. Figure 6.2(b) plots the contact pressure in a three-dimensional coordinate system, with a cone-like distribution.

According to the generalized Iwan model, the contact pressure distribution function is the premise for the derivation of the constitutive force–displacement relationship. Therefore, to simplify the modeling process, the simulated contact pressure distribution is approximated using a linear function. This approximation is widely accepted [2–4]. Figures 6.1(b)–(d) compare the simulated and approximated pressure distribution and shows the relative errors between the simulation and approximation. The relative error is defined as $(p_{FE} - p_{ap})/p_{ap}$, where $p_{FE}$ represents the simulated contact pressure, and $p_{ap}$ the approximated value. It can be seen that the relative errors are almost less than 10% at all locations. Therefore, this approximation is reliable. Additionally, the bolt preload has almost no effect on the distribution of contact pressure.

The approximated contact pressure distribution function can be written as

$$p(r) = p_0\left(1 - \frac{r}{a}\right), \tag{6.1}$$

where the $p_0$ denotes the maximum contact pressure at contact center, $r$ the distance from contact position to center and $a$ the contact radius. The relationship between the contact pressure and the bolt preload is expressed as

$$N = \int_0^a p(r)2\pi r \mathrm{d}r. \tag{6.2}$$

Substituting equation (6.2) into (6.1) yields $p_0 = 3N_0/\pi a^2$.

## 6.2 Modeling friction hysteresis

According to the modeling method of the generalized Iwan model given in chapter 5, the Iwan density function (IDF) and the force–displacement formula are derived from the above pressure distribution function.

### 6.2.1 Sliding stress distribution

The Iwan model is one-dimensional, while the contact pressure on the bolted joint surface is distributed in a two-dimensional spatial domain. Therefore, to relate the two and describe the contact pressure in the Jenkins element coordinates, it is necessary to convert the two-dimensional distribution of contact pressure into an equivalent one-dimensional distribution. To achieve this equivalent conversion, the contact pressure function is integrated along the $y$-axis, as shown in figure 6.3, resulting in the normal load per contact width,

$$p_{eq}(x) = 2\int_0^{\sqrt{a^2 - x^2}} p_0\left(1 - \frac{\sqrt{x^2 + y^2}}{a}\right)\mathrm{d}y$$

$$= p_0\sqrt{a^2 - x^2} + \frac{p_0 x^2}{a}\ln\left(\frac{|x|}{a + \sqrt{a^2 - x^2}}\right). \tag{6.3}$$

Figure 6.3 illustrates the dimensionality reduction of the contact pressure distribution function. In this process, the two-dimensional circular contact area is reduced to

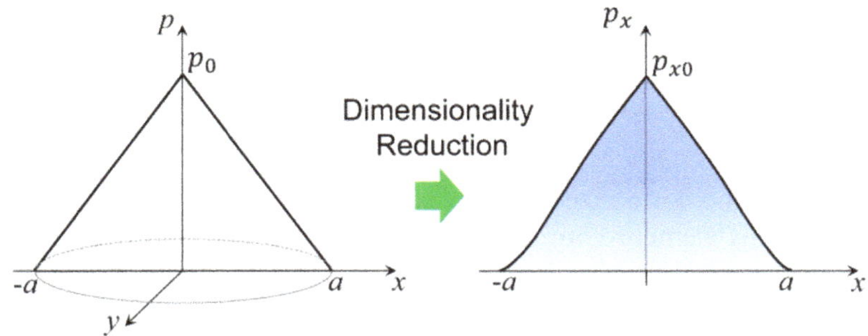


**Figure 6.3.** Dimensionality reduction of pressure distribution from two dimensions to one dimension through an integral process. (Adapted with permission from [1]. Copyright 2020 Springer Nature.)

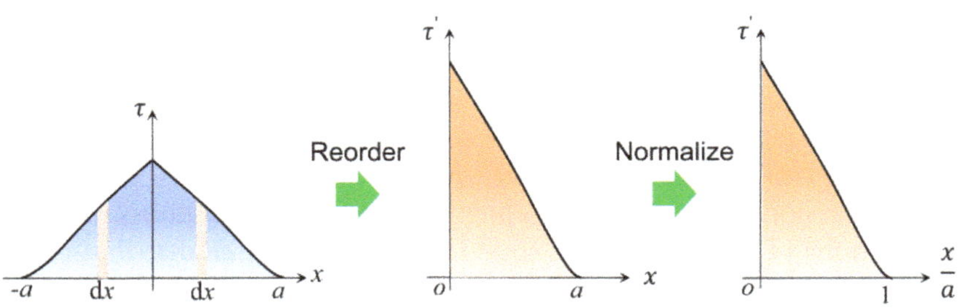


**Figure 6.4.** Reordering the sliding stress distribution and normalizing the contact width for matching the contact coordinates with the number of Jenkins elements. (Adapted with permission from [1]. Copyright 2020 Springer Nature.)

a one-dimensional contact region ($-a \leqslant x \leqslant a$). The normal load per width at the contact center after the dimensionality reduction equals $p_0 a$. By applying Coulomb friction law, the normal contact is related to the tangential sliding friction, resulting in the distribution function of the tangential sliding stress,

$$\tau(x) = \mu p_{\mathrm{eq}}(x), \quad -a \leqslant x \leqslant a. \tag{6.4}$$

Since the distribution function of the sliding stress is symmetric, the function is reconstructed to simplify the model derivation process. As illustrated in figure 6.4, the contact segments with the same sliding stress are stacked together, during which the total sliding stress remains unchanged. The distribution function of the sliding stress after reordering is expressed as

$$\tau'(x) = 2\mu p_0 \sqrt{a^2 - x^2} + 2\frac{\mu p_0 x^2}{a} \ln\left(\frac{x}{a + \sqrt{a^2 - x^2}}\right), \quad 0 \leqslant x \leqslant a. \tag{6.5}$$

Then, to relate the tangential sliding stress distribution function to the IDF, the spatial contact coordinates are normalized. This allows the sliding stress in the spatial contact coordinates to be mapped onto the Jenkins elements. The normalized sliding stress is given by

$$\tau'\left(\frac{x}{a}\right) = 2\mu p_0 a\left\{\sqrt{1-\left(\frac{x}{a}\right)^2} + \left(\frac{x}{a}\right)^2 \ln\left[\frac{\frac{x}{a}}{1+\sqrt{1-\left(\frac{x}{a}\right)^2}}\right]\right\},\quad 0 \leqslant \frac{x}{a} \leqslant 1, \tag{6.6}$$

where $x/a$ denotes the normalized contact coordinates, which can be replaced by the normalized Jenkins element coordinates $i/n$. Substituting $i/n$ into equation (6.6) to replace $x/a$ and multiplying by contact width $a$ yields the distribution function of critical sliding force,

$$f^*\left(\frac{i}{n}\right) = a\cdot\tau'\left(\frac{i}{n}\right) = 2\mu p_0 a^2\left\{\sqrt{1-\left(\frac{i}{n}\right)^2} + \left(\frac{i}{n}\right)^2 \ln\left[\frac{\frac{i}{n}}{1+\sqrt{1-\left(\frac{i}{n}\right)^2}}\right]\right\}. \tag{6.7}$$

According to the definition of the IDF, it is first necessary to obtain the analytical expression of the Jenkins element coordinate $i/n$ with respect to $f^*$. Then, differentiating this expression, the IDF can be derived. However, due to the coupling of the exponential and quadratic terms in equation (6.7), it is difficult to obtain an analytical expression for the Jenkins element coordinates. To address this issue, a set of basis functions $\{1, i/n, (i/n)^2\}$ is used to approximate the distribution function of the critical sliding force. This approximation function $f_a^*(i/n)$ is constrained by following two conditions: (i) the value at the endpoints remains unchanged, i.e. $f_a^*(0) = f^*(0)$ and $f_a^*(1) = f^*(1)$; (ii) the total critical sliding force remains the same as the original, i.e.

$$\int_0^1 f_a^*\left(\frac{i}{n}\right)\mathrm{d}\left(\frac{i}{n}\right) = \int_0^1 f^*\left(\frac{i}{n}\right)\mathrm{d}\left(\frac{i}{n}\right) = \mu N_0. \tag{6.8}$$

Under the constraints of the two aforementioned conditions, the approximation function has a unique expression,

$$f_a^*\left(\frac{i}{n}\right) = 2\mu p_0 a^2\left[(3-\pi)\left(\frac{i}{n}\right)^2 + (\pi-4)\frac{i}{n} + 1\right]. \tag{6.9}$$

Of course, other forms of basis functions or higher-order basis functions are also applicable, which will not be discussed here. Figure 6.5 plots the dimensionless critical sliding stress $f^*/\mu p_0 a^2$ versus normalized contact width and compares the curves before and after the approximation.

### 6.2.2 Iwan density function

The IDF is defined as the cotangent of the tangential sliding stress distribution function, as stated in chapter 5. From equation (6.9), the analytical expression for the Jenkins element coordinates can be computed. Then by differentiating this expression, the IDF can be obtained as

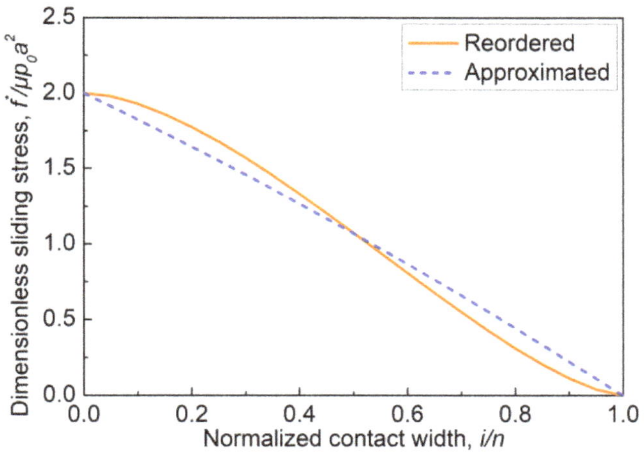


**Figure 6.5.** Dimensionless critical sliding force as a function of normalized contact width. (Adapted with permission from [1]. Copyright 2020 Springer Nature.)

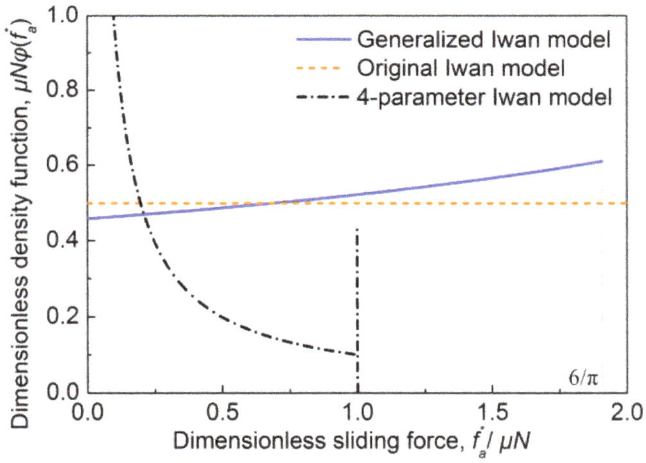


**Figure 6.6.** Dimensionless density function versus dimensionless sliding force. (Adapted with permission from [1]. Copyright 2020 Springer Nature.)

$$\varphi\left(f_a^*\right) = \left|\frac{\mathrm{d}}{\mathrm{d}f_a^*}\left(\frac{i}{n}\right)\right| = \frac{\pi}{6\mu N_0\sqrt{\frac{\pi(6-2\pi)f_a^*}{3\mu N_0} + (\pi - 2)^2}}. \tag{6.10}$$

Figure 6.6 plots the dimensionless IDF $\mu N\varphi(f_a^*)$ as a function of the dimensionless sliding force $f_a^*/\mu N$, and compares the IDF of different Iwan models. Unlike the original Iwan model, the IDF of the generalized Iwan model is a nonlinear function of the critical sliding force. In fact, if a linear basis function $\{1, i/n\}$ is employed to approximate equation (6.7), the resulting IDF is consistent with the original Iwan model. Therefore, the original Iwan model can be regarded as a lower-order form of the generalized Iwan model.

### 6.2.3 Force–displacement relationship

Substituting equation (6.10) into equation (5.2) yields the friction formula under monotonic loading,

$$T(\delta) = \frac{1}{6(\pi - 3)^2}\left\{\left[2(\pi - 3)k_t\delta - \frac{3(\pi - 2)^2\mu N_0}{\pi}\right]\sqrt{\frac{\pi(6 - 2\pi)k_t\delta}{3\mu N_0} + (\pi - 2)^2} + \frac{3(\pi - 2)^3\mu N_0}{\pi} + (3\pi^2 - 21\pi + 36)k_t\delta\right\}. \tag{6.11}$$

Accordingly, substituting equation (6.11) into the Masing formula yields the expression for the hysteresis curve under cyclic loading,

$$\begin{aligned} T_{\text{rel}}(\delta) &= -T(\delta_m) + 2T\left(\frac{\delta_m + \delta}{2}\right),\ \dot{\delta} > 0 \\ T_{\text{unl}}(\delta) &= T(\delta_m) - 2T\left(\frac{\delta_m - \delta}{2}\right),\ \dot{\delta} < 0, \end{aligned} \tag{6.12}$$

where $\delta_m$ is the amplitude of tangential relative displacement. Figure 6.7 plots the normalized friction force as a function of tangential relative displacement under different displacement amplitudes.

The area enclosed by the hysteresis curve represents the friction-induced energy dissipation per cycle and can be calculated by

$$W = \int_{-\delta_m}^{\delta_m} [T_{\text{rel}}(\delta) - T_{\text{unl}}(\delta)]\text{d}\delta. \tag{6.13}$$

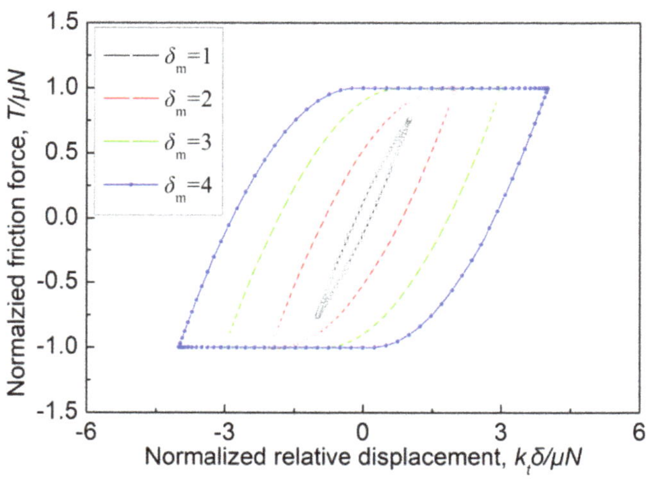


**Figure 6.7.** Evolution of normalized hysteresis curves with displacement amplitudes. (Adapted with permission from [1]. Copyright 2020 Springer Nature.)

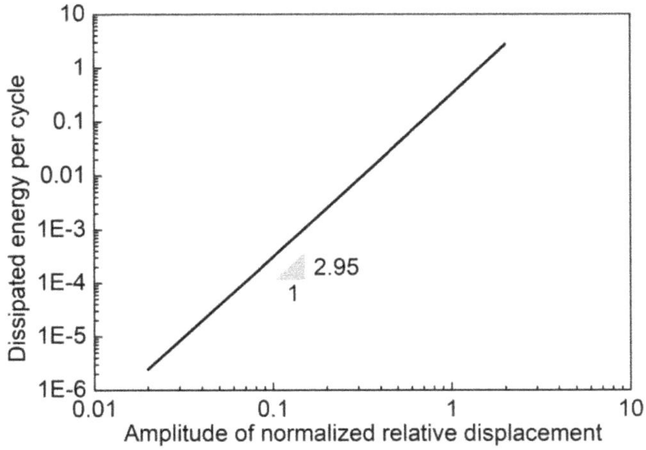


**Figure 6.8.** Dissipated energy per cycle as a function of amplitude of normalized relative displacement. (Adapted with permission from [1]. Copyright 2020 Springer Nature.)

Figure 6.8 shows the dissipated energy per cycle versus the amplitude of relative displacement. In the double logarithmic coordinates, the slope of the curve is about 2.95, which falls within the range [2.2, 3] of experimental findings.

## 6.3 Quasi-static experimental verification

In this section, the generalized Iwan model is used to solve a quasi-static friction problem of bolted connection. The simulation results are compared with experimental counterparts in references [5, 6] for model verification. In addition, comparisons with the original Iwan model and the four-parameter Iwan model are also performed.

Eriten *et al* [5] set up an experimental device to measure fretting friction behavior of bolted connections, as illustrated in figure 6.9(a). The test specimen is pretightened two M3 bolts, with the nominal contact area being a rectangular region $10 \times 17\ \text{mm}^2$ without bolt holes, as shown in figure 6.9(b). Under the action of a piezoelectric actuator, relative motion occurs between the moving and fixed specimens. The relative displacement at the interfaces is measured by a laser nanometer sensor, while the friction force transmitted across the connection surface is measured by a three-dimensional load cell. This set-up allows for direct measurement of the friction force–displacement data of the contact surfaces, avoiding errors introduced by post-processing.

Four sets of experimental results measured under different bolt preloads (234, 331, 526 and 721 N) are selected from [6] and compared with the simulations. Table 6.1 lists the applied displacement amplitudes and contact parameters (including friction coefficient and tangential contact stiffness) for different preloads.

Figure 6.10 presents a comparison between the simulated hysteresis curves and the experimental results. The results show that the simulation results of the generalized Iwan model match well with the experimental results. The difference

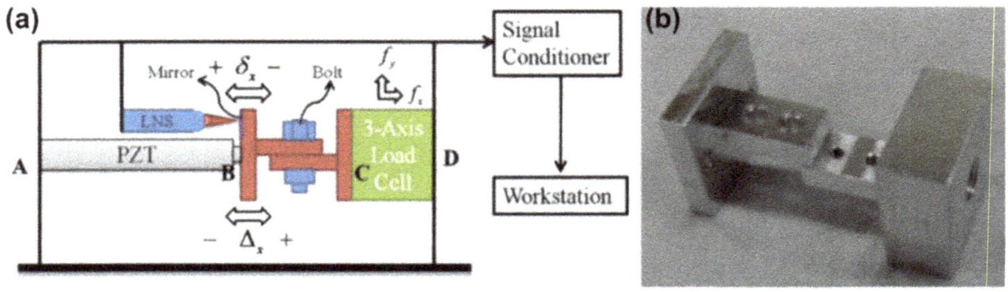


**Figure 6.9.** (a) Sketch of the fretting test apparatus developed by Eriten *et al.* (b) Bolt connection test specimen. (Adapted with permission from [5]. Copyright 2011 Springer Nature.)

**Table 6.1.** Amplitudes of the applied motion and contact parameters for different bolt preloads. (Reproduced with permission from [1]. Copyright 2020 Springer Nature.)

| Bolt preload (N) | 234 | 331 | 526 | 721 |
|---|---|---|---|---|
| Max. tangential displacement ($\mu$m) | 4.16 | 2.62 | 1.25 | 0.71 |
| Friction coefficient | 0.479 | 0.508 | 0.420 | 0.323 |
| Tangential stiffness (N $\mu$m$^{-1}$) | 122.7 | 160.2 | 277.7 | 491.6 |

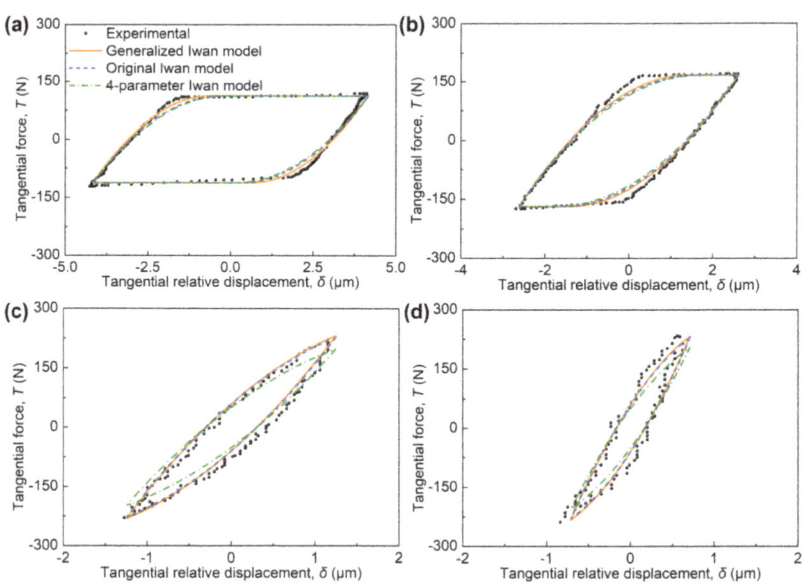


**Figure 6.10.** Comparison between the simulated and measured force–displacement curves under different bolt preloads: (a) $N = 234$ N, (b) $N = 331$ N, (c) $N = 526$ N, and (d) $N = 721$ N. (Adapted with permission from [1]. Copyright 2020 Springer Nature.)

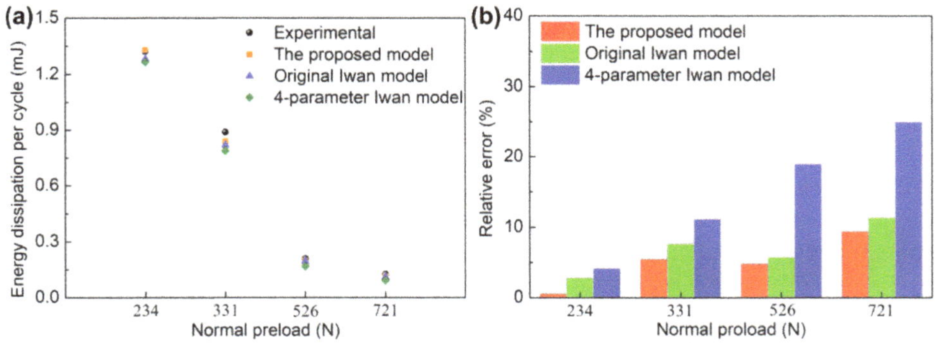


**Figure 6.11.** Comparison between the simulated and measured energy dissipation: (a) dissipated energy per cycle at different bolt preloads and (b) relative errors of predicted energy dissipation. (Adapted with permission from [1]. Copyright 2020 Springer Nature.)

between the simulation results of the original Iwan model and the generalized Iwan model is minimal, indicating that the higher-order terms have little impact on the results. In contrast, the difference between the simulated curves by the four-parameter Iwan model and the experimental results is more pronounced. From the first two cases, it can be observed that in the early stages of motion at the bolted connection interface, the four-parameter Iwan model quickly enters the partial slip state, while the generalized Iwan model undergoes a relatively longer stick phase. During the transition from the partial slip state to the gross slip state, the generalized Iwan model enters the gross slip stage first. For the latter two cases, under the same applied displacement, the friction force predicted by the four-parameter Iwan model is more conservative compared to the generalized Iwan model. By comparing the shapes of the hysteresis loops, it can be seen that the hysteresis loop predicted by the generalized Iwan model exhibits higher stiffness than that predicted by the four-parameter Iwan model.

Figure 6.11(a) presents the energy dissipation per cycle under four different conditions. The comparison between the simulations and the experimental results indicates that the generalized Iwan model has a higher accuracy in predicting energy dissipation compared to the four-parameter Iwan model. Figure 6.11(b) illustrates the relative errors between the predicted energy dissipation and the experimental counterparts. For the first case, the relative error of the generalized Iwan model is very small, less than 1%. For the remaining three cases, the error is relatively higher but still below 10%. The relative error of the original Iwan model is slightly higher than that of the generalized Iwan model but remains acceptable. Compared to the generalized Iwan model, the four-parameter Iwan model exhibits higher relative errors across all four conditions, exceeding 20% in some cases.

The good consistency between the simulation results and the experimental results verifies the effectiveness of the generalized Iwan model in bolted connections. Compared to the traditional four-parameter Iwan model, the generalized Iwan model can better predict the hysteresis curves and energy dissipation. Therefore, it can be concluded that the generalized Iwan model can accurately simulate the friction hysteresis behavior of bolted connection interfaces. Moreover, the IDF based on contact pressure distribution is more reliable than the assumed one.

## 6.4 Dynamics prediction of a bolted joint structure

In this section, the generalized Iwan model is used to simulate the friction nonlinear behavior of the bolted connection interface in a bolted joint oscillator system, demonstrating its applicability in the dynamic analysis of connected structures. Due to the relatively simple mathematical formulation of this model, it is fully compatible with the latest numerical analysis tools. The dynamic response is calculated using the harmonic balance method (HBM) [7–9] and the alternating frequency/time domain method (AFT) [10–12].

### 6.4.1 The Gaul resonator

Figure 6.12 shows the schematic diagram of the bolted joint oscillator system made of stainless steel, consisting of a frame on the left side and a lumped mass on the right side [13, 14]. The left frame includes a flexible component that ensures the desired vibration mode and resonance frequency of the device. The left frame and the right mass block are connected by an M10 bolt. The nominal contact area is a $40 \times 40$ mm$^2$ square surrounding the through hole. The frame and the mass block are horizontally suspended on a fixed bracket to avoid potential misalignment along the horizontal direction. A sinusoidal sweep excitation is applied to the left frame by an electromagnetic exciter and transmitted along the central axis of the device.

The acceleration of each component is captured by accelerometers, and the corresponding displacement response is obtained through frequency-domain integration. The nonlinear friction force transmitted across the bolted connection interface is captured by the inertia force of the right mass block. The hysteresis curve at the connection interface is characterized by the relative displacement $u_2$–$u_B$ and the inertia force of the right mass block, where $u_2$ represents the displacement of

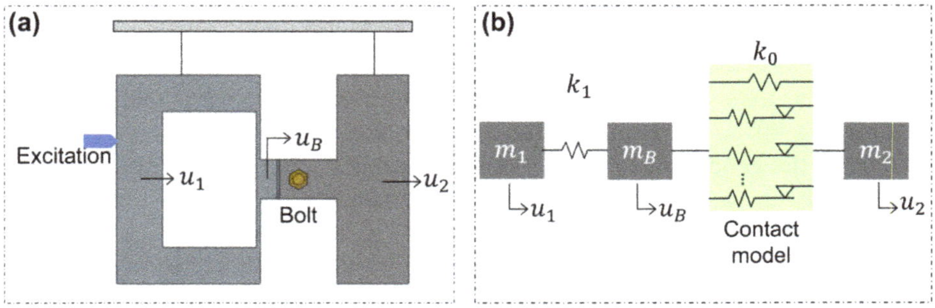


**Figure 6.12.** (a) Schematic diagram of the bolt connection vibration test device [14] and (b) corresponding three degrees of freedom model in which the joint interface is simulated by the contact model. (Adapted with permission from [1]. Copyright 2020 Springer Nature.)

**Table 6.2.** Model parameters of the bolted joint oscillator and contact parameters. (Reproduced with permission from [1]. Copyright 2020 Springer Nature.)

| $m_1$ (kg) | $m_2$ (kg) | $m_B$ (kg) | $k_t$ (N $\mu$m$^{-1}$) | $k_0$ (N $\mu$m$^{-1}$) | $k_1$ (N $\mu$m$^{-1}$) | $\mu$ |
|---|---|---|---|---|---|---|
| 5.304 | 5.199 | 0.54 | 691.0 | 121.0 | 12.5 | 0.533 |

the right mass and $u_B$ the displacement of the central part of the device, as shown in figure 6.12(a).

Figure 6.12(b) illustrates a simplified mechanical model of the bolted joint oscillator system, which has three degrees of freedom. The symbols $m_1$, $m_B$, $m_2$ represent the masses of each part, $u_1$ the displacement of the left frame, $k_1$ the stiffness of the flexible component, and $k_0$ the residual stiffness of the connection interface. These parameters are listed in table 6.2.

The dynamic equilibrium equation of the bolted joint oscillator system is written as

$$\boldsymbol{M}\ddot{\boldsymbol{u}}(t) + \boldsymbol{C}\dot{\boldsymbol{u}}(t) + \boldsymbol{K}\boldsymbol{u}(t) + \boldsymbol{f}_{\mathrm{nl}}(\boldsymbol{u}, \dot{\boldsymbol{u}}, t) = \boldsymbol{F}_e(t), \tag{6.14}$$

in which $\boldsymbol{M}$, $\boldsymbol{C}$, $\boldsymbol{K}$ are mass, damping, and stiffness matrices, respectively; $\boldsymbol{f}_{\mathrm{nl}}$ is nonlinear force vector which depends on displacement and velocity; $\boldsymbol{F}_e$ is excitation force vector. The specific form of equation (6.14) is written as

$$\begin{bmatrix} m_1 & 0 & 0 \\ 0 & m_B & 0 \\ 0 & 0 & m_2 \end{bmatrix} \begin{Bmatrix} \ddot{u}_1 \\ \ddot{u}_B \\ \ddot{u}_2 \end{Bmatrix} + \begin{bmatrix} k_1 & -k_1 & 0 \\ -k_1 & k_1 + k_0 & -k_0 \\ 0 & -k_0 & k_0 \end{bmatrix} \begin{Bmatrix} u_1 \\ u_B \\ u_2 \end{Bmatrix} + \begin{Bmatrix} 0 \\ T \\ -T \end{Bmatrix} = \begin{Bmatrix} F_e \\ 0 \\ 0 \end{Bmatrix}. \tag{6.15}$$

In the following simulations, the effect of material damping is neglected because friction damping dominates the structural damping of this device.

### 6.4.2 Dynamic response calculation method

Since friction force is non-continuous and nonlinear, equation (6.15) cannot be solved analytically in the frequency domain. Here, the HBM and the AFT method are used to calculate the dynamic response of the bolted joint oscillator. This method combines the efficiency of solving algebraic equations in the frequency domain with the convenience of solving nonlinear friction forces in the time domain, resulting in relatively high computational efficiency.

The HBM approximates the dynamic response using Fourier series. Taking the displacement response as an example, the actual displacement solution can be expanded into a series of harmonic components,

$$\boldsymbol{u} = \boldsymbol{U}_o + \sum_{j=1}^{n_h} \left[ \boldsymbol{U}_j^c \cos(j\omega t) + \boldsymbol{U}_j^s \sin(j\omega t) \right], \tag{6.16}$$

where $\boldsymbol{U}_j^c$ and $\boldsymbol{U}_j^s$ are the cosinoidal and sinusoidal harmonic coefficients, respectively, $\boldsymbol{U}_o$ the constant component of displacement, $j$ the order of Fourier series, $n_h$ the truncated order of harmonics and $\omega$ the principal vibration frequency.

Similarly, the vectors of nonlinear force and excitation force can also be approximated in the same way,

$$\boldsymbol{f}_{\text{nl}} = \boldsymbol{f}_o + \sum_{j=1}^{n_h}\left[\boldsymbol{f}_j^c \cos(j\omega t) + \boldsymbol{f}_j^s \sin(j\omega t)\right], \tag{6.17}$$

$$\boldsymbol{F}_e = \boldsymbol{F}_o + \sum_{j=1}^{n_h}\left[\boldsymbol{F}_j^c \cos(j\omega t) + \boldsymbol{F}_j^s \sin(j\omega t)\right]. \tag{6.18}$$

The dynamics governing equation (6.14) can be converted into the form of algebraic equation,

$$\boldsymbol{Z}(\omega)\boldsymbol{u} = \boldsymbol{F}_e - \boldsymbol{f}_{\text{nl}}, \tag{6.19}$$

in which $\boldsymbol{Z}(\omega)$ is the dynamic stiffness matrix of the linear part of the system and defined as

$$\boldsymbol{Z}(\omega) = \text{diag}(\boldsymbol{K}, \boldsymbol{Z}_1, \ldots, \boldsymbol{Z}_j, \ldots, \boldsymbol{Z}_{n_h}), \tag{6.20}$$

where $\boldsymbol{Z}_j$ $(j = 1, 2, \ldots, n_h)$ is defined as

$$\boldsymbol{Z}_j = \begin{bmatrix} -(j\omega)^2\boldsymbol{M} + \boldsymbol{K} & j\omega\boldsymbol{C} \\ -j\omega\boldsymbol{C} & -(j\omega)^2\boldsymbol{M} + \boldsymbol{K} \end{bmatrix}. \tag{6.21}$$

Equation (6.19) is usually rewritten as a residual form,

$$\boldsymbol{R} = \boldsymbol{Z}(\omega)\boldsymbol{u} - \boldsymbol{F}_e + \boldsymbol{f}_{\text{nl}}. \tag{6.22}$$

where $\boldsymbol{R}$ is the residual. The residual equation can be solved using the Newton–Raphson iteration method for the steady-state response of the system,

$$\boldsymbol{u}^{(j+1)} = \boldsymbol{u}^{(j)} - \left[\left.\frac{\partial \boldsymbol{R}}{\partial \boldsymbol{u}}\right|_{\boldsymbol{u}^{(j)}}\right]^{-1} \boldsymbol{R}(\boldsymbol{u}^{(j)}), \tag{6.23}$$

where $\partial\boldsymbol{R}/\partial\boldsymbol{u}$ is the Jacobian matrix of the system,

$$\frac{\partial \boldsymbol{R}}{\partial \boldsymbol{u}} = \boldsymbol{Z}(\omega) + \frac{\partial \boldsymbol{f}_{nl}}{\partial \boldsymbol{u}}. \tag{6.24}$$

Since friction force depends on the relative displacement and velocity between the contact surfaces, and their relationship is nonlinear and hysteretic, it is difficult to calculate the friction force directly in the frequency domain. Therefore, the AFT method is used to convert the response signal between the time domain and frequency domain, allowing the friction force to be calculated in the time domain. Figure 6.13 illustrates the computational process of the AFT method.

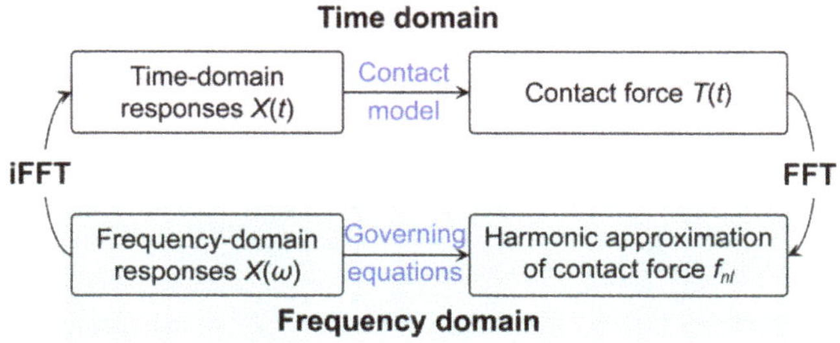


**Figure 6.13.** Schematic of the alternating time-frequency method (AFT).

The specific computational steps of the AFT method are as follows: First, the initial response $\boldsymbol{u}(t)$ of the system in time domain are obtained without considering the nonlinear contact force. Then, the contact force $f_{\mathrm{nl}}(t)$ in time domain is calculated based on the contact model and converted to the frequency domain through a fast Fourier transform (FFT). Subsequently, the obtained frequency-domain contact force $\boldsymbol{f}_{\mathrm{nl}}(\omega)$ is substituted into equation (6.19), and the algebraic equation is iteratively solved to obtain a new frequency-domain system response $\boldsymbol{u}(\omega)^{(i+1)}$. Finally, this frequency-domain response is then converted back to the time domain through an inverse FFT (iFFT), and the friction force is updated. This process is repeated until the solution converges, with the convergence criterion being that the residual is below a given tolerance.

### 6.4.3 Dynamic response

The authors of [14] experimentally studied the frequency response of the aforementioned bolted joint oscillator under different excitation amplitudes (12.5, 25, and 50 N). The bolt preload was set to 1.5 kN. This section uses the HBM and the AFT method to solve the frequency response curves of the bolted joint oscillator, and investigates the impact of truncated harmonic orders on the results.

Figures 6.14(a), (c), and (e) illustrate the experimental frequency response curves and simulation results of the right lumped mass block under the three sets of excitation amplitudes. The results show that the frequency response curves exhibit typical nonlinear characteristics, specifically, as the excitation amplitude increases, the resonance peak of the system shifts towards the lower frequency range (stiffness softening), and the resonance response peak gradually decreases. The comparison between the predictions and experimental results demonstrates that the simulation method has good predictive performance for the dynamic response of bolted joint structures.

Figures 6.14(b), (d), and (f) show the effects of truncated harmonic orders on the frequency response curves. The results show that the differences between the frequency response curves for different harmonic orders are mainly concentrated around the resonance peak, and the maximum relative error between them is less than 5%. It indicates that under the aforementioned conditions, the impact of harmonic orders on the results is minimal and can even be considered negligible.

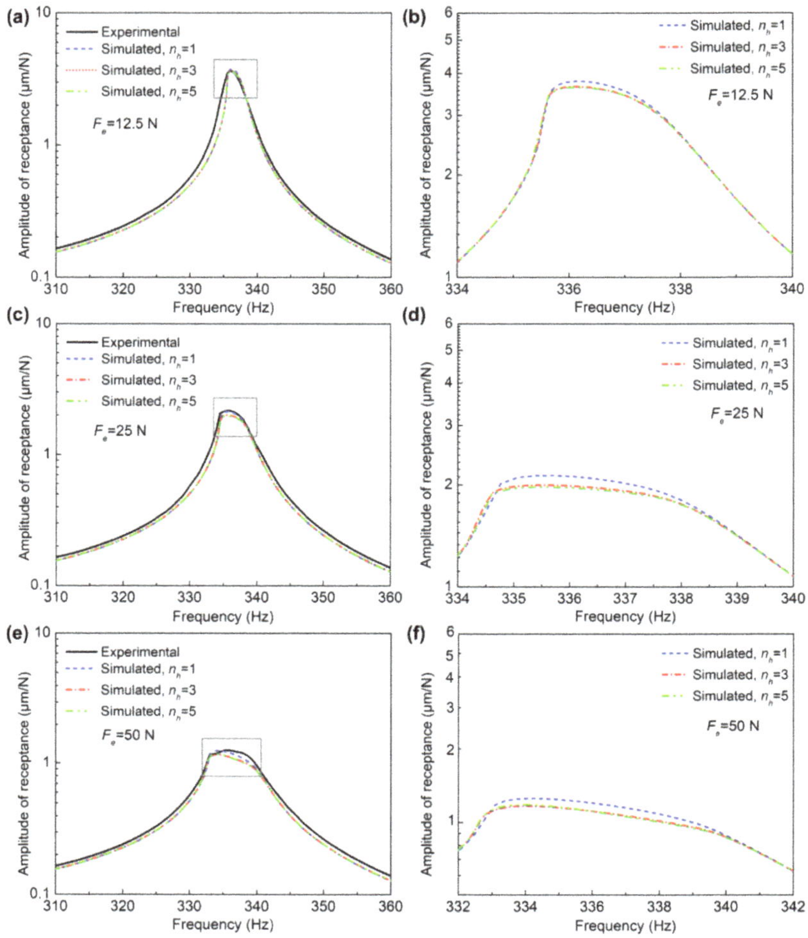


**Figure 6.14.** Comparison between predicted and measured frequency response curves for different excitation amplitudes: (a)–(b) $F_e$ = 12.5 N, (c)–(d) $F_e$ = 25 N, and (e)–(f) $F_e$ = 50 N.

In addition, figure 6.15 plots the hysteresis loops near the resonance frequency under an excitation amplitude of 50 N. The hysteresis curve at 336 Hz in figure 6.15 (a) exhibits gross slip, while in figure 6.15(b) the curve always remains in the partial slip regime. Overall, the simulation predictions align well with the experimental results. Nevertheless, there are still some visible differences between the two, which may be primarily attributed to the following reasons. (i) Contact parameters might change with the loading direction, such as residual stiffness. The slope of in the hysteresis loop during the gross slip phase in figure 6.15(a) does not match the slope of the measured curve, which could be due to unstable contact parameters caused by changes in contact conditions during the testing process. Additionally, the tangential contact stiffness during the reloading and unloading stages are also slightly different. (ii) Another major reason could be the inaccuracy of the measured bolt preload or

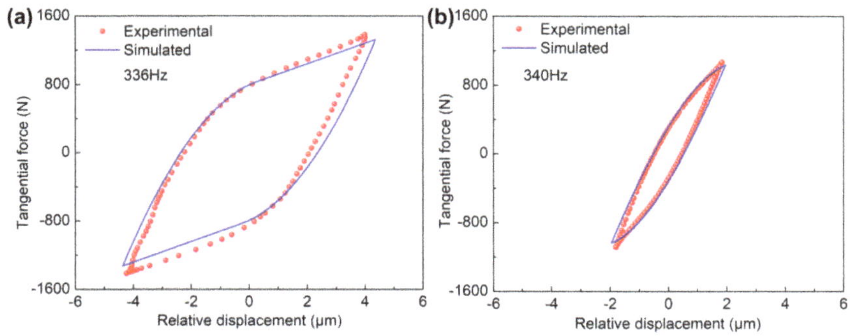


**Figure 6.15.** Comparison between predicted and measured hysteresis loops at two frequencies: (a) $f = 336$ Hz and (b) $f = 340$ Hz. (Adapted with permission from [1]. Copyright 2020 Springer Nature.)

the estimated friction coefficient, as the predicted tangential friction force is lower than the experimental results. Other possible factors include the redistribution of contact pressure at the interface, additional structural damping, and the effects of fretting wear.

## 6.5 Summary

This chapter applies the generalized Iwan model to characterize the friction contact at bolted joint interfaces and verifies the model accuracy under both quasi-static and dynamic loading. The main content and conclusions are as follows:

1. Based on finite element contact analysis, the contact pressure distribution at bolted joint interfaces is obtained. A mapping relationship between the contact pressure distribution function and the IDF is established, and the force–displacement expression for the friction hysteresis of the joint interface is derived.
2. The generalized Iwan model is used to solve the quasi-static friction problem of a typical bolted connection. The results show that the generalized Iwan model can reproduce the experimental results well and predict the nonlinear hysteresis and energy dissipation better than the those traditional models.
3. The generalized Iwan model is applied to the dynamic response analysis of a multi-degree-of-freedom bolted joint oscillator. The steady-state response of the system is solved using the harmonic balance method and the alternating frequency/time method. The results validate the effectiveness of the dynamic modeling and demonstrate the compatibility of the interface model with the nonlinear dynamic analysis codes.

## Bibliography

[1] Li D, Xu C, Kang J and Zhang Z 2020 Modeling tangential friction based on contact pressure distribution for predicting dynamic responses of bolted joint structures *Nonlinear Dyn.* **101** 255–69

[2] Kim J, Yoon J C and Kang B S 2007 Finite element analysis and modeling of structure with bolted joints *Appl. Math. Modell.* **31** 895–911
[3] Marshall M B, Lewis R and Dwyer-Joyce R S 2006 Characterisation of contact pressure distribution in bolted joints *Strain* **42** 31–43
[4] Nelson N R, Prasad N S and Sekhar A S 2023 Structural integrity and sealing behaviour of bolted flange joint: a state of art review *Int. J. Press. Vessels Pip.* **204** 104975
[5] Eriten M, Polycarpou A A and Bergman L A 2011 Development of a lap joint fretting apparatus *Exp. Mech.* **51** 1405–19
[6] Eriten M 2012 *Multiscale Physics-based Modeling of Friction* (Champaign, IL: University of Illinois at Urbana-Champaign)
[7] Krack M and Gross J 2019 *Harmonic Balance for Nonlinear Vibration Problems* (Cham: Springer International)
[8] Sanliturk K Y and Ewins D J 1996 Modelling two-dimensional friction contact and its application using harmonic balance method *J. Sound Vib.* **193** 511–23
[9] Firrone C M, Zucca S and Gola M M 2011 The effect of underplatform dampers on the forced response of bladed disks by a coupled static/dynamic harmonic balance method *Int. J. Non Linear Mech.* **46** 363–75
[10] Cameron T M and Griffin J H 1989 An alternating frequency/time domain method for calculating the steady-state response of nonlinear dynamic systems *ASME J. Appl. Mech.* **56** 149–54
[11] Prabith K and Krishna I R P 2020 The numerical modeling of rotor–stator rubbing in rotating machinery: a comprehensive review *Nonlinear Dyn.* **101** 1317–63
[12] Lacayo R, Pesaresi L, Groß J, Fochler D, Armand J, Salles L, Schwingshackl C, Allen M and Brake M 2019 Nonlinear modeling of structures with bolted joints: a comparison of two approaches based on a time-domain and frequency-domain solver *Mech. Syst. Sig. Process.* **114** 413–38
[13] Gaul L, Lenz J and Sachau D 1998 Active damping of space structures by contact pressure control in joints *J. Struct. Mech.* **26** 81–100
[14] Süß D 2016 *Multi-Harmonische-Balance-Methoden zur Untersuchung des Übertragungsverhaltens von Strukturen mit Fügestellen* (Erlangen: Friedrich-Alexander-Universität Erlangen-Nürnberg (FAU))

**IOP** Publishing



**Chao Xu and Dongwu Li**

# Chapter 7

# A physics-based multi-scale friction model

This chapter introduces a multi-scale frictional modeling method that comprehensively considers the random roughness characteristics of the interface, contact pressure distribution, and the convenience of a phenomenological model. In particular, the model parameters do not need to be identified through dynamic testing or precise contact measurements. In addition, a novel fretting test device based on a fatigue testing machine is presented. The novelty of this experimental device lies in the symmetrical arrangement design to eliminate the influence of residual stiffness, and the use of flexible components to prevent additional torque from affecting the contact area.

## 7.1 The multi-scale modeling approach

As mentioned in chapter 5, there are two main modeling approaches to simulate frictional contact behavior: models that consider rough contact (physics-based models) [1–7] and phenomenological contact models (data-driven models) [8–13]. Recently, a multi-scale fusion modeling approach was developed, combining the multi-scale features of randomly rough surfaces and the convenience of phenomenological contact models in friction hysteresis simulation [14–17].

The physics-based models take into account the rough surface properties and physics-based constitutive relations, but are unlikely to be compatible with structural dynamics analysis programs because the resulting force–displacement expressions are not explicit [17]. As a comparison, the data-driven models show good accuracy in the simulation of friction hysteresis and have good compatibility with structural dynamics analysis. However, the estimation procedure of the model parameters is not convenient for predicting the dynamic response of connected structures. Generally, this procedure requires dedicated fretting tests or dynamic frequency-response tests [18–23].

This section introduces a multi-scale fusion friction modeling method to calculate the tangential contact stiffness and reproduce the friction hysteresis of bolted joint

doi:10.1088/978-0-7503-6214-6ch7  

interfaces, which comprehensively considers the rough contact, the non-uniform distribution of contact pressure, and the phenomenological contact model [17]. The tangential contact stiffness is calculated using the fractal contact theory and the friction hysteresis is reproduced by the discrete Iwan model, with the former providing the necessary model parameters for the latter.

### 7.1.1 Contact pressure

Fractal contact theory implies an important assumption that the contact pressure in the planar contact problem is uniformly distributed, which does not apply to bolted connection interfaces at which the contact pressure monotonically decreases along the contact radius. Clearly, areas near the contact edge are more susceptible to sliding than areas near the contact center. Therefore, when using the fractal contact theory to simulate joint interfaces, the non-uniform pressure distribution must be considered. Li *et al* [17] divided the contact area into several sections and assumed that the contact pressure in the subdivided local areas is constant, where fractal contact theory can be applied.

Similar to chapter 6, an approximated function of the contact pressure distribution at bolted joint interfaces is assumed that the contact pressure decreases linearly along the contact radius and remains unchanged in the circumferential direction [24, 25], as shown in figure 7.1.

The contact pressure function $p(r)$ can be written as

$$p(r) = p_0\left(\frac{R_{\max} - r}{R_{\max} - R_{\min}}\right), \tag{7.1}$$

where $R_{\min}$ and $R_{\max}$ denote the radius of the bolt hole and the contact radius, respectively, $r$ the distance to the hole center, and $p_0$ the maximum pressure. The relationship between the bolt preload $N_0$ and the contact pressure function satisfies

$$N_0 = \int_{R_{\min}}^{R_{\max}} 2\pi r p(r)\mathrm{d}r. \tag{7.2}$$

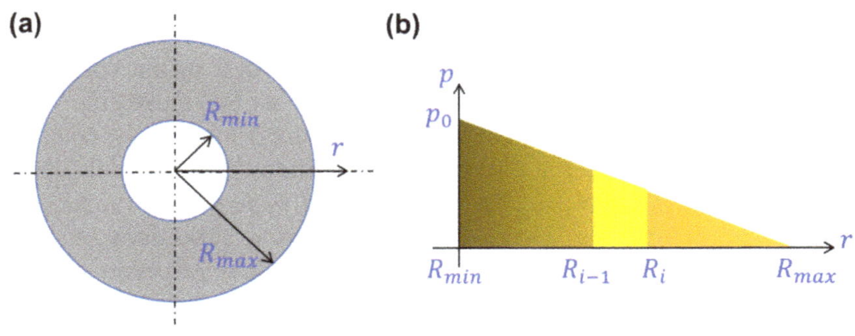


**Figure 7.1.** (a) Contact area of the bolted joint interface and (b) approximated contact pressure distribution along the contact radius. (Adapted with permission from [17]. Copyright 2022 Elsevier.)

Substituting equation (7.1) into equation (7.2) gives the maximum contact pressure,

$$p_0 = \frac{3N_0}{\pi\left(R_{\max}^2 + R_{\max}R_{\min} - 2R_{\min}^2\right)}. \tag{7.3}$$

The friction state (stick or sliding) of connected interfaces depends on the contact pressure distribution. In order to consider the influence of contact pressure on friction contact, the authors of [17] divide the contact area into $h$ annular subregions along the contact radius. Each annular region has the same radial length, $(R_{\max} - R_{\min})/h$. The normal load $N_{0i}$ of the $i$th annular region can be calculated,

$$N_{0i} = \int_{R_{i-1}}^{R_i} 2\pi r p(r)\mathrm{d}r = \frac{2\pi p_0}{R_{\max} - R_{\min}}\left[\frac{R_{\max}(R_i^2 - R_{i-1}^2)}{2} - \frac{(R_i^3 - R_{i-1}^3)}{3}\right], \tag{7.4}$$

where the upper and lower limits of the integral are the boundaries of the $i$th annular region, as shown in figure 7.1(b),

$$R_{i-1} = R_{\min} + \frac{i-1}{h}(R_{\max} - R_{\min}), \tag{7.5}$$

$$R_i = R_{\min} + \frac{i}{h}(R_{\max} - R_{\min}). \tag{7.6}$$

In each of the divided annular regions, the contact pressure is assumed to be uniformly distributed, allowing fractal contact theory to be applied to analyze the contact area of the bolted joint interface, and thereby calculate the tangential contact stiffness.

### 7.1.2 Fractal contact model

According to rough contact theory, the contact between two rough surfaces can be simplified to the contact between a rough surface and a rigid smooth surface. Figure 7.2 shows the simplified contact and the deformation geometry of a single asperity. To consider the multi-scale nature of the rough surface, fractal theory is used to describe the random rough surface.

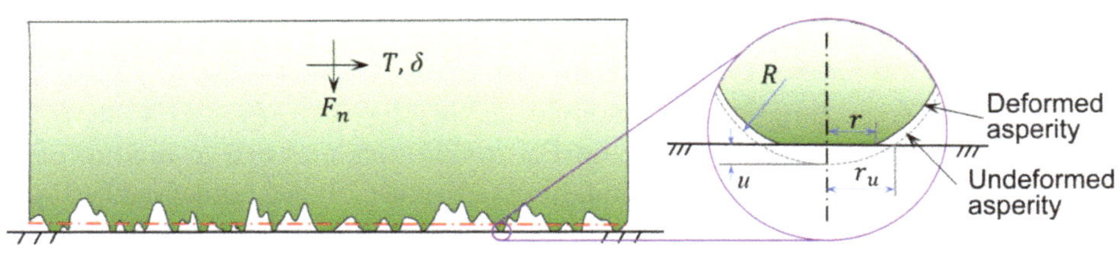


**Figure 7.2.** Schematic diagram of the contact between a random rough surface and a rigid smooth surface and the deformation geometry of an individual asperity. (Adapted with permission from [17]. Copyright 2022 Elsevier.)

The modified Weierstrass–Mandelbrot (W–M) function [26] is widely used to characterize a three-dimensional topography of random rough surfaces. Here, for simplicity, a simplified W–M function is employed to describe the rough surface height $z(x)$,

$$z(x) = L\left(\frac{G}{L}\right)^{D-2}(\ln\gamma)^{0.5}\sum_{n=0}^{n_{\max}}\gamma^{(D-3)n}\left[\cos\phi_{1,n} - \cos\left(\frac{2\pi\gamma^n x}{L} - \phi_{1,n}\right)\right], \tag{7.7}$$

where $L$ denotes the sampling length of the surface profile, $G$ the fractal roughness, $D$ the fractal dimension ($2 < D < 3$), $\gamma$ the scaling factor that is related to the frequency spectrum of the surface profile (typically $\gamma = 1.5$), $n$ the frequency index and $n_{\max}$ corresponding to the high cutoff frequency of the profile, $\phi_{1,n}$ the random phase, and $x$ the coordinate along the surface. The normal deformation of a single asperity can be expressed as

$$u = 2G^{D-2}(\ln\gamma)^{0.5}(2r_u)^{3-D}, \tag{7.8}$$

where $r_u$ denotes the truncated radius of the undeformed asperity, as shown in the zoomed-in view in figure 7.2.

According to the deformation geometry of an individual asperity, $R^2 - (R-u)^2 = r_u^2$ and $R \gg u$, the curvature radius $R$ of the asperity can be derived as

$$R = 2^{D-5}G^{2-D}(\ln\gamma)^{-0.5}\left(\frac{a_u}{\pi}\right)^{0.5D-0.5}, \tag{7.9}$$

where $a_u$ is the truncated area of the undeformed asperity and equals $\pi r_u^2$.

### 7.1.3 Normal contact

Under a given normal load, asperities of different heights on a rough surface generally exhibit three deformation states: purely elastic, elastic–plastic, and fully plastic, denoted by the subscripts $e$, $ep$, and $p$, respectively. Consequently, each asperity experiences different normal loads, deformation, and contact areas.

1. *Purely elastic contact*

   For asperities with relatively low heights, their deformation is small and they exhibit purely elastic behavior. According to the Hertz contact theory [27], the contact force $F_e$ acting on these asperities can be written as

$$F_e = \frac{4}{3}E^*R^{0.5}u^{1.5}, \tag{7.10}$$

   where $E^*$ is the reduced elastic modulus, $E^* = [(1-\vartheta_1^2)/E_1 + (1-\vartheta_2^2)/E_2]^{-1}$ with $\vartheta_1$, $E_1$, $\vartheta_2$, and $E_2$ being the Poisson's ratios and elastic moduli of two contacting bodies. Substituting equations (7.8) and (7.9) into equation (7.10) yields the contact force $F_e$ as a function of the truncated contact area $a_u$,

$$F_e(a_u) = \frac{2^{5.5-D}}{3\pi^{2-0.5D}}E^*G^{D-2}(\ln\gamma)^{0.5}a_u^{2-0.5D}. \tag{7.11}$$

2. *Fully plastic contact*

For relatively high asperities, they undergo fully plastic deformation. The contact force $F_p$ can be obtained according to [14],

$$F_p(a_u) = 2\pi RHu = Ha_u, \tag{7.12}$$

where $H$ is the hardness of the softer material.

3. *Elastic–plastic contact*

With a moderate asperity height, the asperity transitions from purely elastic deformation to fully plastic deformation. According to [28], the contact force induced by the elastoplastic deformation of the asperity is written as

$$F_{ep} = k_\vartheta H\pi R u_{ec}{}^{-\beta} u^{1+\beta}, \tag{7.13}$$

in which $k_\vartheta$ represents the mean contact pressure factor, $u_{ec}$ the critical deformation for the onset of plastic yield, and $\beta = \ln(2/k_\vartheta)/\ln(u_{pc}/u_{ec})$, where $u_{pc}$ denotes the critical deformation at the beginning of fully plastic deformation and is equal to $110u_{ec}$ [28]. Following [29], $k_\vartheta$ can be expressed as a function of Poisson's ratio,

$$k_\vartheta = 0.3097 + 0.2094\vartheta + 0.1295\vartheta^2. \tag{7.14}$$

The critical deformation $u_{ec}$ at onset of plastic yield is given, based on [30],

$$u_{ec} = \left(\frac{0.75\pi k_\vartheta H}{E^*}\right)^2 R. \tag{7.15}$$

Then, substituting equations (7.8), (7.9), and (7.15) into equation (7.13) gives the expression of the normal force $F_{ep}$ as a function of the truncated contact area $a_u$,

$$F_{ep}(a_u) = 2^{13\beta-2\beta D-1}\left(\frac{E^*}{3}\right)^{2\beta}\pi^{\beta(D-4)}\left(k_\vartheta H\right)^{1-2\beta}G^{2\beta(D-2)}\left(\ln\gamma\right)^{\beta}a_u{}^{1+2\beta-\beta D}. \tag{7.16}$$

The reason for expressing the contact force as a function of the truncated area is that the total contact force $F_n$ can be obtained by integrating the product of the contact force of a single asperity and the area distribution function $n(a_u)$, as follows,

$$F_n = \begin{cases} \int_0^{a_{u\_p}} F_p n(a_u)\mathrm{d}a_u + \int_{a_{u\_p}}^{a_{u\_e}} F_{ep} n(a_u)\mathrm{d}a_u + \int_{a_{u\_e}}^{a_{u\_l}} F_e n(a_u)\mathrm{d}a_u, & a_{u\_l} > a_{u\_e} \\ \int_0^{a_{u\_p}} F_p n(a_u)\mathrm{d}a_u + \int_{a_{u\_p}}^{a_{u\_l}} F_{ep} n(a_u)\mathrm{d}a_u, & a_{u\_p} < a_{u\_l} < a_{u\_e} \\ \int_0^{a_{u\_l}} F_p n(a_u)\mathrm{d}a_u, & a_{u\_l} < a_{u\_p} \end{cases}, \tag{7.17}$$

where the integral limits $a_{u\_p} = 2\pi R u_{pc}$, $a_{u\_e} = 2\pi R u_{ec}$, and $a_{u\_l}$ represents the maximum truncated area. According to [26], the area distribution function $n(a_u)$ is given as

$$n(a_u) = \frac{D-1}{2} a_{u\_l}^{\frac{D-1}{2}} a_u^{\frac{-D-1}{2}}. \tag{7.18}$$

The maximum truncated area $a_{u\_l}$ of a single asperity can be expressed as a function of the real contact area $A_r$, following [4],

$$a_{u\_l} = 2\frac{3-D}{D-1} A_r. \tag{7.19}$$

A detailed formula of the normal force $F_n$ is given as follows. If $a_{u\_l} > a_{u\_e}$, the normal load is given as

$$F_n = \int_0^{a_{u\_p}} F_p n(a_u) \mathrm{d}a_u + \int_{a_{u\_p}}^{a_{u\_e}} F_{ep} n(a_u) \mathrm{d}a_u + \int_{a_{u\_e}}^{a_{u\_l}} F_e n(a_u) \mathrm{d}a_u. \tag{7.20}$$

The first term in equation (7.20) is the normal force of those asperities under fully plastic deformation, which is derived as

$$F_{n\_p} = \frac{\eta \sigma_y (D-1)}{3-D} a_{u\_l}^{\frac{D-1}{2}} a_{u\_p}^{\frac{3-D}{2}}, \tag{7.21}$$

where $\sigma_y$ denotes yield strength of the softer material and $\eta$ a constant approximately equal to 2.79. The second term represents the normal force caused by the elastic–plastic deformation of asperities and equals

$$F_{n\_ep} = \left\{ \begin{array}{l} \dfrac{2^{13\beta-2\beta D-2} \pi^{\beta(D-4)}}{1.5+2\beta-\beta D-0.5D} (k_\vartheta H)^{1-2\beta} G^{2\beta(D-2)} (\ln\gamma)^\beta (D-1) a_{u\_l}^{\frac{D-1}{2}} \\ \times \left(\dfrac{E^*}{3}\right)^{2\beta} (a_{u\_e}^{1.5+2\beta-\beta D-0.5D} - a_{u\_p}^{1.5+2\beta-\beta D-0.5D}) \end{array} \right\}. \tag{7.22}$$

The third term is the normal force of those asperities under the purely elastic deformation,

$$F_{n\_e} = \frac{2^{4.5-D} E^* G^{D-2} (\ln\gamma)^{0.5} (D-1)}{(3\pi^{2-0.5D})(2.5-D)} a_{u\_l}^{0.5D-0.5} (a_{u\_l}^{2.5-D} - a_{u\_e}^{2.5-D}). \tag{7.23}$$

If $a_{u\_p} < a_{u\_l} < a_{u\_e}$, the normal load is written as

$$\begin{aligned} F_n &= \int_0^{a_{u\_p}} F_p n(a_u) \mathrm{d}a_u + \int_{a_{u\_p}}^{a_{u\_l}} F_{ep} n(a_u) \mathrm{d}a_u \\ &= F_{n\_p} + \frac{2^{13\beta-2\beta D-2} \pi^{\beta(D-4)}}{1.5+2\beta-\beta D-0.5D} (k_\vartheta H)^{1-2\beta} G^{2\beta(D-2)} (\ln\gamma)^\beta (D-1) a_{u\_l}^{0.5D-0.5} \left(\frac{E^*}{3}\right)^{2\beta} \\ &\times \left(a_{u\_e}^{1.5+2\beta-\beta D-0.5D} - a_{u\_p}^{1.5+2\beta-\beta D-0.5D}\right). \end{aligned} \tag{7.24}$$

If $a_{u\_l} < a_{u\_p}$, the normal load is

$$F_n = \int_0^{a_{u\_l}} F_p n(a_u) \mathrm{d}a_u = \frac{\eta\sigma_y(D-1)a_{u\_l}}{3-D}. \tag{7.25}$$

Once the fractal and material parameters of the contacting bodies are known, the actual contact area can be obtained by setting the normal force $F_n$, equation (7.17), equal to the applied normal load $N_{0i}$, equation (7.6), in the $i$th annular region of the bolted joint interface. Then, the actual contact area $A_{ri}$ of the $i$th annular contact region is used as a known parameter to solve for the tangential contact.

### 7.1.4 Tangential contact stiffness

Similarly, starting from the tangential contact of a single asperity, the tangential contact stiffness $k_{ti}$ of the $i$th annular contact region can be determined by integrating the individual contact stiffness with the area distribution function $n(a_u)$. Note that the 'tangential contact stiffness' here corresponds to the interface stick regime and does not change with the movement of contact interfaces. As mentioned in chapter 5, Mindlin [31] gave the friction solution for the contact between two elastic spheres. The tangential deformation of the sphere under a monotonic load is written as

$$\delta = \frac{3\mu F_n}{16G^*r}\left[1-\left(1-\frac{T}{\mu F_n}\right)^{2/3}\right], \tag{7.26}$$

where $\mu$, $G^*$, $r$, and $T$ denote the friction coefficient, reduced shear modulus, contact radius, and tangential load, respectively. $G^*=[(2-\vartheta_1)/G_1+(2-\vartheta_2)/G_2]^{-1}$, where $G_1$ and $G_2$ are the shear modulus of two contacting bodies. The tangential contact stiffness of a single asperity can be obtained by taking the limit of the derivative $\partial T/\partial\delta$ at $\delta \to 0$,

$$k_{t\_\mathrm{asp}} = \lim_{u\to 0}\frac{\partial T}{\partial \delta} = \lim_{u\to 0} 8G^*r\left(1-\frac{16G^*r\delta}{3\mu F_n}\right)^{1/2} = 8G^*r. \tag{7.27}$$

Based on the Hertz contact theory and the deformation characteristics of a single asperity, the contact radius can be obtained, $r=\sqrt{a_u/2\pi}$. Substituting this formula into equation (7.27) yields the relationship between the tangential stiffness and the truncated area,

$$k_{t\_\mathrm{asp}} = 4\sqrt{\frac{2}{\pi}}G^*{a_u}^{0.5}. \tag{7.28}$$

Finally, the tangential contact stiffness $k_{ti}$ of the $i$th annular contact region is determined by integrating the product of $k_{t\_\mathrm{asp}}$ and the area distribution function,

$$k_{ti} = \int_{a_{u\_e}}^{a_{u\_l}} k_{t\_\mathrm{asp}} n(a_u)\mathrm{d}a_u = \frac{4\sqrt{2}\,G^*(D-1)\psi^{1.5-0.5D}}{\sqrt{\pi}(2-D){a_{u\_l}}^{0.5-0.5D}}\left({a_{u\_l}}^{1-0.5D}-{a_{u\_e}}^{1-0.5D}\right). \tag{7.29}$$

### 7.1.5 Friction hysteresis

For the simulation of friction hysteresis, a discretized Iwan model was used. It comprehensively considers the non-uniform distribution of contact pressure and the surface rough characteristics, as depicted in figure 7.3.

Unlike the original Iwan model, the discretized Iwan model consists of several Jenkins elements connected in parallel, each with different stiffness $k_{ti}$ and critical sliding force $f_i = \mu N_{0i}$, as shown in figures 7.3(b) and (c). The stiffness $k_{ti}$ is determined by equation (7.29), while the critical sliding force depends on the distribution function of normal loads, equation (7.4).

For monotonic tangential loading, the force–displacement relationship of Jenkins element can be expressed as

$$T_i = \begin{cases} k_{ti}\delta, & 0 < \delta \leqslant \dfrac{f_i}{k_{ti}}. \\ f_i, & \text{else} \end{cases} \tag{7.30}$$

For cyclic loading, the friction force can be determined as

$$T_i = \begin{cases} k_{ti}(\delta - \delta_0) + f_i, & \delta_0 \geqslant \dfrac{f_i}{k_{ti}}, \ \dot{\delta} < 0, \ \delta_0 - \dfrac{2f_i}{k_{ti}} < \delta \leqslant \delta_0 \\ -f_i, & \delta_0 \geqslant \dfrac{f_i}{k_{ti}}, \ \dot{\delta} < 0, \ -\delta_0 < \delta \leqslant \delta_0 - \dfrac{2f_i}{k_{ti}} \\ k_{ti}(\delta + \delta_0) - f_i, & \delta_0 \geqslant \frac{f_i}{k_{ti}}, \ \dot{\delta} > 0, \ -\delta_0 \leqslant \delta < \frac{2f_i}{k_{ti}} - \delta_0, \\ f_i, & \delta_0 \geqslant \frac{f_i}{k_{ti}}, \ \dot{\delta} > 0, \ \frac{2f_i}{k_{ti}} - \delta_0 \leqslant \delta < \delta_0 \\ k_{ti}\delta, & \delta_0 < \frac{f_i}{k_{ti}} \end{cases} \tag{7.31}$$

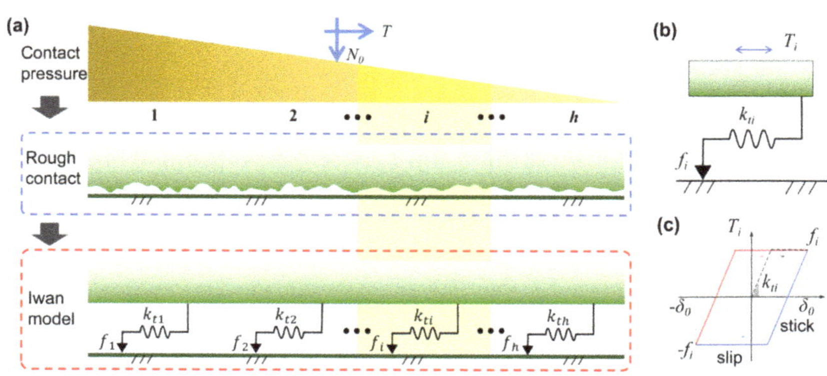


**Figure 7.3.** (a) Schematic diagram of modeling the friction hysteresis behavior by considering the contact pressure distribution and rough contact in the Iwan model, (b) Jenkins element and (c) force–displacement curve under cyclic tangential loads. (Adapted with permission from [17]. Copyright 2022 Elsevier.)

where $\dot{\delta}$ denotes the velocity and $\delta_0$ the amplitude of displacement. The total friction force can be obtained by

$$T = \sum_{i=1}^{i=h} T_i. \tag{7.32}$$

## 7.2 Experimental tests

To validate the accuracy of the above multi-scale friction model, the authors of [17] developed a new fretting test apparatus of bolted connection. This device is specifically designed to avoid the influence of residual stiffness and minimize the impact of additional bending moments at the connection interface when contact conditions are reached. This section introduces the fretting test set-up and typical experimental results.

### 7.2.1 Experimental method

Figure 7.4 shows an overview and a photograph of the test set-up. Unlike the test rig in chapter 2, the bolted connection sample in this set-up, which has two moving specimens and a fixed specimen, is arranged in a symmetrical layout to capture the friction force only acting on the bolted joint interface. As explained in chapter 2, in an asymmetrical layout, the friction forces at the interfaces (between the screw head, washer, nut, and target interface) and the bending deformation force of bolt shank are also included in the measured force, which leads to errors in the friction measurement, as shown in figure 7.5. These factors also originate of residual stiffness in the gross slip regime. In this experiment, a symmetrical specimen layout was adopted to avoid the influence of them on the friction measurement.

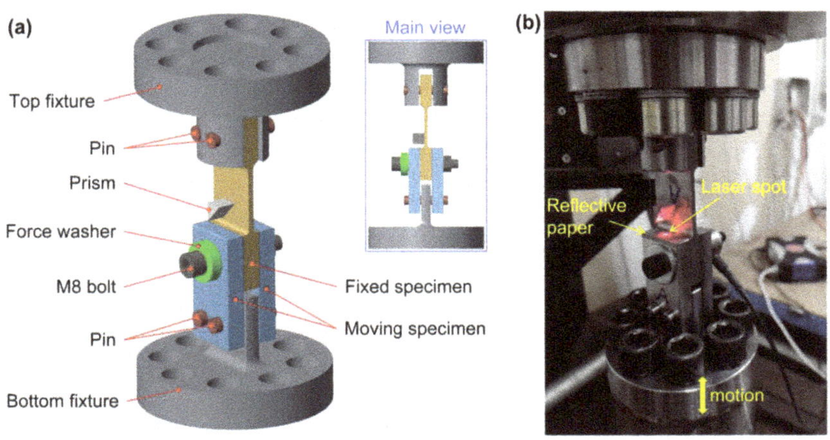


**Figure 7.4.** (a) Schematic diagram of the fretting test apparatus and main components. (b) Photograph of the servo-hydraulic fatigue testing system and the assembled bolted joint specimens. (Adapted with permission from [17]. Copyright 2022 Elsevier.)

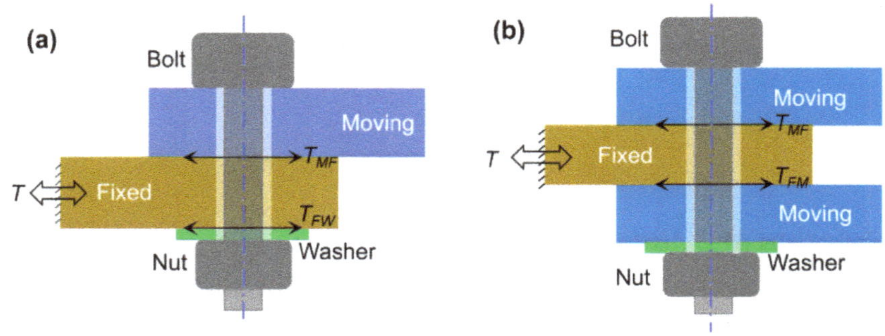


**Figure 7.5.** (a) Asymmetric layout of bolt joints used in [32] in which the measured friction force includes that at other surfaces, not only that at the target surface. (b) Symmetric layout in [17] where the measured friction force is equal to the resultant of those at two target surfaces. (Adapted with permission from [17]. Copyright 2022 Elsevier.)

The specimens are made of ASTM 304 stainless steel and are fastened together with an M8 bolt. There are two parallel contact areas, each contact region of a 30 × 30 $mm^2$ square with a 10 mm diameter hole in the center. The moving specimens are connected to the bottom fixture by two 8 mm diameter pins. There is a 1 mm gap between both sides of the protruding part in the middle of the bottom fixture and the moving specimen to avoid contact other than the target, see the main view in figure 7.4(a). The fixed specimen is specially designed to have greater flexibility along the normal direction of the contact surface, thereby avoiding as much as possible the tilt contact (or local contact which is found from wear scars in some tests [33]) between the moving and fixed surfaces caused by the extra moment during installation. The fixed specimen looks like a dumbbell overall. This design resembles the 'floating body approach' in [34]. From this perspective, the thinner the leaf spring, the more conducive it is to forming a planar contact. However, if the leaf spring is too thin, the test device will be at risk of buckling instability under a large compression load. Therefore, considering both flexibility and structural stability, the thickness of the leaf spring is set to 3 mm. Therefore, considering both flexibility and structural stability, the thickness of the leaf spring is set to 3 mm. The other end of the fixed specimen is connected to the top fixture through two pins with a diameter of 8 mm. Similarly, a 1 mm gap is maintained between the top fixture and the fixed specimen.

The top and bottom fixtures are mounted on a servo-hydraulic fatigue testing system (capacity of 250 kN, Instron Inc.). The upper end of the top fixture and the lower end of the bottom fixture both have a cylindrical recess designed to fit with the protrusions on the piston shaft, ensuring that the fixture axis aligns as closely as possible with the piston shaft. The piston shaft drives the motion specimen to reciprocate, causing relative movement at the bolted joint interfaces. As shown in figure 7.5(b), the friction force measured by a built-in load cell of the fatigue testing system should be the resultant of the tangential contact forces of the two surfaces. The relative displacement measurement is carried out in the same way as in chapter

2, using a laser vibrometer (OFV-525, Polytec Inc.) in combination with a small prism (size 10 × 10 × 10 mm) with an inclination of 45°. The prism is mounted on the fixed specimen near the contact area and is used to reflect the laser beam onto the end of the moving specimen pasted with reflective paper, as shown in figure 7.4(b).

It is worth noting that because the bolted joint specimen is not strictly constrained along the normal direction of the contact surface, the piston excitation may cause slight vibrations in the specimen along the normal direction. Therefore, to eliminate the impact of this factor on the displacement measurement accuracy, the laser beam path is set to be parallel to the contact surface, as shown in figure 7.4(b).

In addition, due to the compliance of the leaf spring, the relative motion between the prism and the ideal measuring point on the fixed specimen should not be ignored when measuring the relative displacement of the target interface. To quantify this compliance and further correct the measured displacements, a finite element model of the fixed specimen is established, as shown in figure 7.6 where the relative displacement between the prism and the ideal measurement position is plotted versus the compressive load. It shows that the compliance of the leaf spring between the prism and the ideal measurement position is 0.34 $\mu$m kN$^{-1}$. Therefore, the measured relative displacements are corrected by subtracting the product of the compliance and measured force.

The bolt preload is recorded by an annular force washer (KMR-40 kN for M8 bolts, HBM Inc.). Before each test, the target surfaces are cleaned with acetone and measured using an Alicona Infinite Focus instrument. All tests were carried at room temperature.

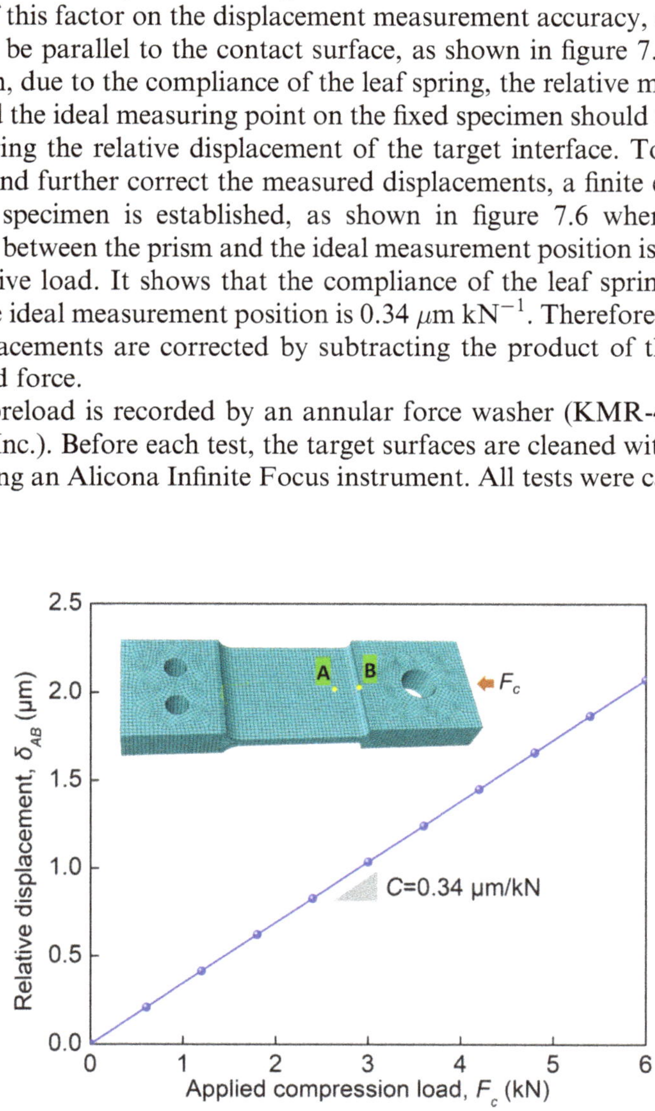


**Figure 7.6.** Relative displacement between points A and B as a function of the applied compression load simulated by the finite element method. Points A and B represent laser spots on the prism and the ideal measurement position on the fixed specimen, respectively. (Adapted with permission from [17]. Copyright 2022 Elsevier.)

### 7.2.2 Experimental results

Figure 7.7 compares a hysteresis loop measured by this experimental set-up with that obtained in [32] where the joint specimen is arranged in an asymmetric layout and fastened with an M6 bolt. For a more intuitive comparison, the original hysteresis loops are normalized by dividing by the amplitude of the force/displacement. The contact surface in [32] sequentially experiences stick, partial slip, and gross slip.

In contrast, the curve in [17] shows a complex shape after the initial stick state, with three distinct decreases in tangential contact stiffness. The first decrease occurs at zero friction force, where the friction force almost does not change with the increase in interface relative displacement. This indicates the presence of gaps in the load transmission path of the experimental set-up, most likely at the pin connections between the fixture and the specimen. Although the gap effect temporarily causes a loss of 'measured contact stiffness', it does not affect the energy dissipation in each cycle. The second decrease occurs during the partial slip phase, which may be due to one contact surface undergoing gross slip while the other remains in partial slip. Due to differences in microscopic surface morphology, it is impossible to maintain the same motion state even if the two contact surfaces are symmetrically distributed and loaded simultaneously. The third decrease is caused by gross slip on both contact surfaces. During this phase the tangential contact stiffness drops to zero, which differs from the findings in [32].

Figure 7.8 shows the hysteresis loops measured at different excitation amplitudes. The excitation amplitude is obtained from the piston displacement. The bolt preload

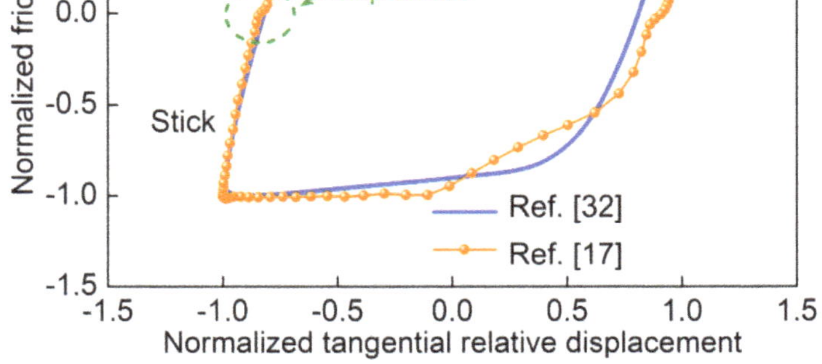


**Figure 7.7.** Comparison of the normalized hysteresis loops measured in [17] with that in [32]. Normalization is implemented by dividing the original measurement data by the corresponding amplitude. (Adapted with permission from [17]. Copyright 2022 Elsevier.)

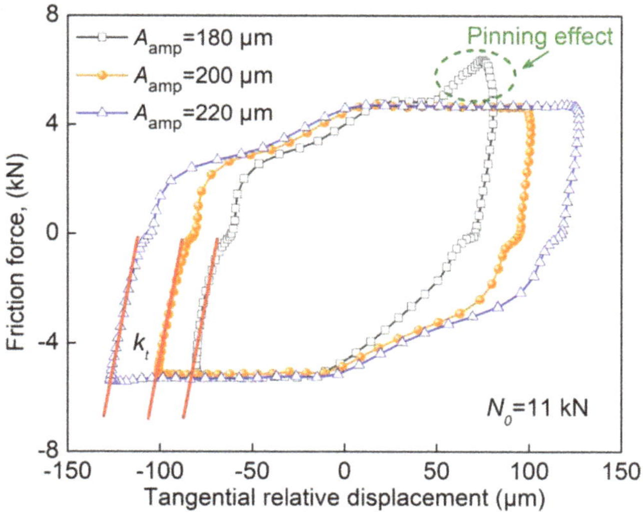


**Figure 7.8.** Friction force versus tangential relative displacement for varying excitation amplitudes and the same bolt preload of 11 kN. (Adapted with permission from [17]. Copyright 2022 Elsevier.)

is set to 11 kN. Although the tangential movement of the connection specimen causes oscillations in the preload, this variation does not exceed ±5% of the initial value throughout the test. The results show that under different excitation amplitudes, the initial tangential contact stiffness (slope of the red solid line) and the friction coefficient remain almost unchanged. At the end of the gross slip stage, an 'uplift' can be observed in the hysteresis curve for the 185 $\mu$m condition. This phenomenon is caused by the contact between the screw and the bolt hole. As explained in chapter 3, when tightening the bolt, it is difficult to control the position of the screw relative to the bolt hole. In some cases, the screw is too close to the inner wall of the bolt hole after being tightened. Once the sliding distance of the contact surface exceeds the gap between the screw and the hole, a new contact is thus formed between them, resulting in the 'uplift' phenomenon. This phenomenon has also been observed in [10, 32].

Figure 7.9 illustrates the variation of the ratio of friction force to bolt preload with respect to tangential relative displacement under different bolt preloads and excitation amplitudes. The shapes of the hysteresis curves for these four scenarios are similar. This ratio in the gross slip regime varies slightly with bolt preload. This variation may be attributed to changes in surface topography caused by multiple tests.

This fretting test apparatus effectively minimizes the impact of contact surfaces other than the target surface on friction force measurements, giving it high accuracy and uniqueness among the fretting test rigs found in the literature to date. Then, the experimental results are used as a benchmark to verify the multi-scale friction model.

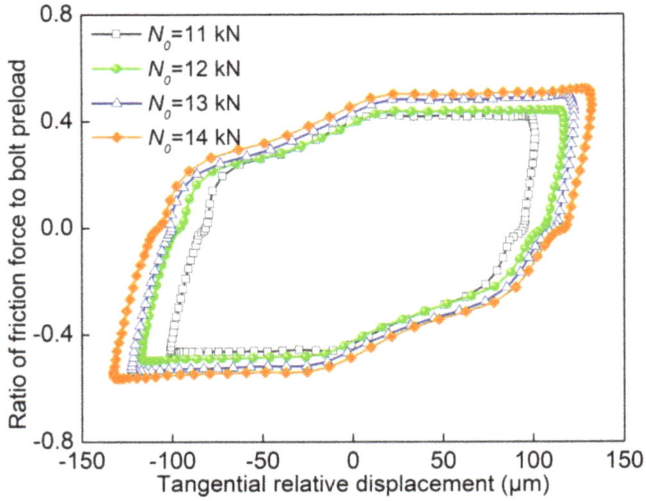


**Figure 7.9.** The ratio of friction force to bolt preload as a function of tangential relative displacement for varying loading cases. (Adapted with permission from [17]. Copyright 2022 Elsevier.)

## 7.3 Model validation and discussion

This section introduces the parameter estimation method of the multi-scale friction model and verifies the effectiveness of the model by comparison with the experimental results in section 7.2. Then, a parametric analysis is given to discuss the effect of the contact radius and the number of the divided annular sub-regions on the simulations.

### 7.3.1 Model parameter estimation

The model parameters include contact radius $R_{\max}$, fractal dimension $D$, fractal roughness $G$, and friction coefficient $\mu$. Generally, the contact radius of joint interfaces can be obtained using direct measurement (based on pressure-sensitive film) and finite element methods. In [35, 36], the authors show that the contact radii obtained by these two methods are very consistent, with a relative error of only 5%. In the finite element model, they assume that the contact surfaces are smooth. Therefore, the authors of [17] use the finite element method to calculate the contact radius at the bolted interface, with all specimens meshed using 8-node linear hexahedron elements. Figure 7.10 plots the contact radius versus bolt preload, showing a monotonic increase trend in the contact radius with bolt preload. The influence of the surface roughness layer of the joint surface on the contact area will be discussed in section 7.3.3.

The fractal dimension $D$ and fractal roughness $G$, are identified using the power spectral density (PSD) function method [37]. The logarithm of the PSD of the surface profile can be expressed as

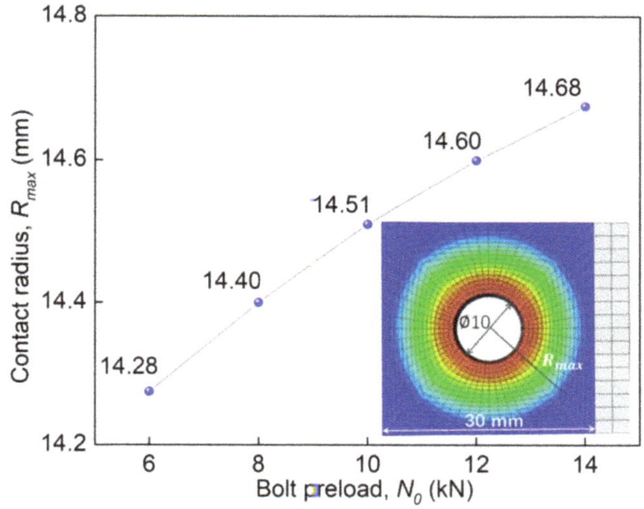


**Figure 7.10.** Simulated contact radius $R_{\max}$ under different bolt preloads by finite element contact analysis. (Reproduced with permission from [17]. Copyright 2022 Elsevier.)

$$\log P_{sd}(\omega) = (2D - 7)\log\omega + (2D - 4)\log G - \log(\ln\gamma) + (6 - 2D)\log\pi + (5 - 2D)\log 2, \tag{7.33}$$

where $P_{sd}$ denotes the PSD and $\omega$ the wavevector of surface profile. The fractal parameters can be identified from the PSD–wavevector curve of the surface profile height according to the following expression,

$$k_{\text{PSD}} = 2D - 7, \tag{7.34}$$

$$b_{\text{PSD}} = (2D - 4)\log G - \log(\ln\gamma) + (6 - 2D)\log\pi + (5 - 2D)\log 2, \tag{7.35}$$

where $k_{\text{PSD}}$ and $b_{\text{PSD}}$ are the slope and intercept of the PSD–wavevector curve, respectively. $k_{\text{PSD}}$ and $b_{\text{PSD}}$ can be directly determined from the PSD–wavevector curve of the surface profile. The surface topography is measured using an optical 3D measurement system (Alicona G4 Infinite Focus). Figure 7.11(a) shows a photograph of the fixed specimen, where the straight yellow lines (FB1-4) indicate the measurement path of the surface profile. Figure 7.11(b) plots the measured surface profile height versus sampling length.

The surface roughness $R_a$ and PSD curve can be obtained from the measured surface topography data, with the average value of the measured roughness taken as the roughness of the contact surface. Figure 7.12 shows an example of the PSD curves of surface profile, where the fitted curve is obtained using the least squares method. The fractal parameters can be calculated through substituting the slope and intercept of the fitted PSD curve into equations (7.34) and (7.35). Table 7.1 lists the surface roughness $R_a$ and fractal parameters $D$ and $G$. Since the materials of the connected specimens are the same and the surface roughness is similar, the average

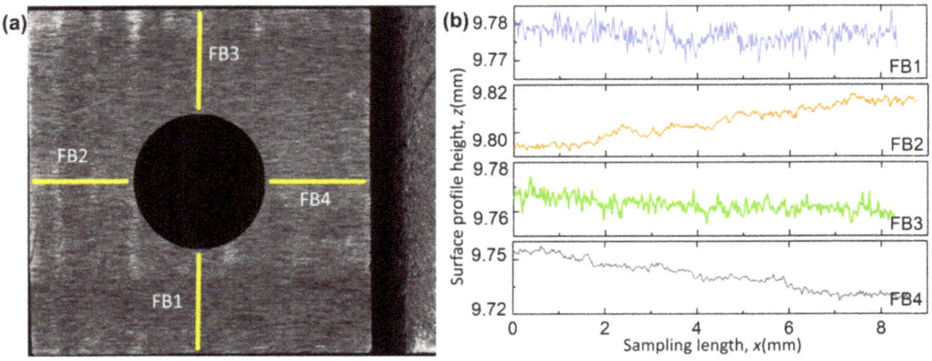


**Figure 7.11.** (a) Measured surface photograph using an optical 3D measurement system. (b) Surface profile height curves at different measurement positions (FB1-4). (Reproduced with permission from [17]. Copyright 2022 Elsevier.)

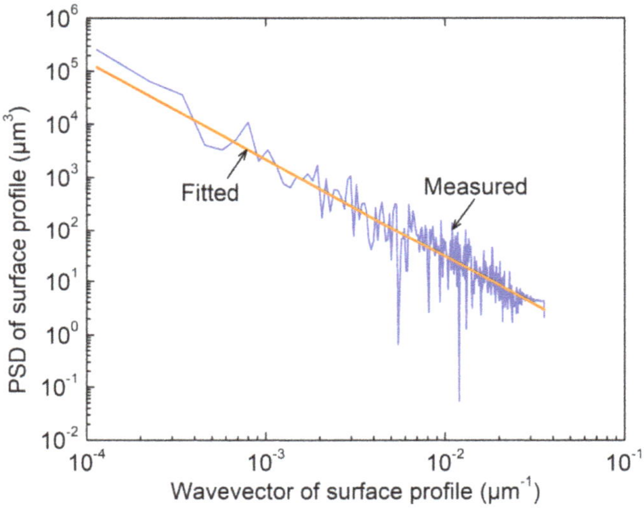


**Figure 7.12.** PSD curve of surface profile height (at FB1) in the double logarithmic coordinate system as a function of wavevector: measurement and least square fitting. (Reproduced with permission from [17]. Copyright 2022 Elsevier.)

**Table 7.1.** Surface roughness and fractal parameters of contact surfaces. Moving_$A_{surf}$, Fixed_$A_{surf}$, Moving_$B_{surf}$, and Fixed_$B_{surf}$ represent the contact surfaces of the joint specimens. Interface_A and Interface_B represent the two target interfaces. (Reproduced with permission from [17]. Copyright 2022 Elsevier.)

| Surface roughness, $R_a$ ($\mu$m) | | Fractal parameters, $D$ and $G$ | | |
|---|---|---|---|---|
| Moving_$A_{surf}$ | 1.28 | Interface_A | $D$ | 2.523 |
| Fixed_$A_{surf}$ | 1.39 | | $G$ (m) | $8.11 \times 10^{-8}$ |
| Moving_$B_{surf}$ | 1.38 | Interface_B | $D$ | 2.588 |
| Fixed_$B_{surf}$ | 1.47 | | $G$ (m) | $3.25 \times 10^{-9}$ |

slope and intercept of all fitted PSD curves can be used to determine the equivalent fractal parameters. Note that fractal parameters cannot be directly averaged.

Another parameter to be identified is the friction coefficient. To accurately predict the frictional hysteresis response, the friction coefficient is directly extracted from the measured hysteresis loops. Additionally, it is assumed that the friction coefficient is the same for both contact interfaces.

### 7.3.2 Model validation

The material parameters of the connected components are as follows: 200 GPa elastic modulus, 0.29 Poisson's ratio, 86 GPa shear modulus, and 203 HB hardness (converted to 687 MPa tensile strength). Four experimental hysteresis curves under different preload conditions are selected for comparison with the simulation results. Additionally, the discretized Iwan models composed of different numbers of Jenkins elements (or divided sub-regions) are also compared.

Figure 7.13 shows the simulated and measured hysteresis curves. The multi-scale friction model (with $h = 4$) can reproduce the measured approximately hexagonal hysteresis curves well. The differences between the simulation results and the measurements mainly occur during the partial slip phase at the joint interface. There are various reasons for this discrepancy. On one hand, after multiple tests, the surface topography may change due to wear effects, potentially leading to deviations

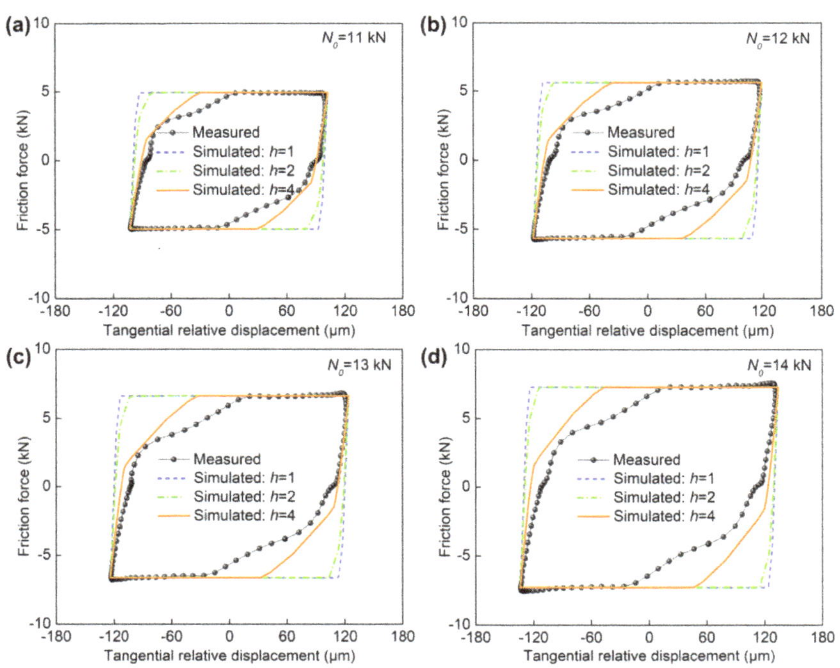


**Figure 7.13.** Comparison of simulated hysteresis curves with experimental curves under different bolt preloads: (a) $N_0 = 11$ kN, (b) $N_0 = 12$ kN, (c) $N_0 = 13$ kN, and (d) $N_0 = 14$ kN. (Reproduced with permission from [17]. Copyright 2022 Elsevier.)

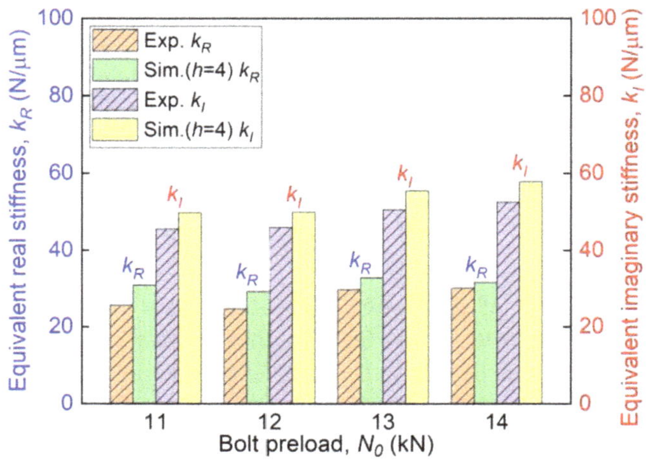


**Figure 7.14.** Quantitative comparison of the simulated equivalent real stiffness and the equivalent imaginary stiffness with the experimental counterparts under different bolt preloads. (Reproduced with permission from [17]. Copyright 2022 Elsevier.)

in the identified fractal parameters. On the other hand, due to surface waviness and the possible additional bending moments, the actual contact pressure distribution at the joint interface may differ from the assumed distribution. Additionally, the contact model does not account for the gap effects observed in the experiment, and there are also differences in the actual friction coefficients of the two contact surfaces. However, considering that the modeling approach is physically based and multi-scale, these differences are acceptable to some extent.

Compared with the two cases of $h = 1$ and $h = 2$, the simulation results (by the model with $h = 4$) are more consistent with the experimental results. The effect of the number of Jenkins elements (or divided sub-regions) on the model prediction capability will be given in section 7.3.4.

To quantitatively assess the accuracy of the proposed modeling method, the equivalent real stiffness $k_R$ (representing elastic stiffness) and imaginary stiffness $k_I$ (representing damping) are obtained from the first-order Fourier expansion of the force–displacement curve [38]. It can be seen from figure 7.14 that the relative differences in comparisons are smaller (almost all below 10%). Overall, the multi-scale modeling method can reproduce the frictional hysteresis behavior at the bolted interface with acceptable accuracy.

### 7.3.3 Effect of contact radius

This subsection first analyzes the sensitivity of the simulated hysteresis loop with respect to the contact radius. Figure 7.15 presents a numerical case in which a 20% dispersion of the contact radius $R_{\max}$ is considered (the gray shade). It shows that the effect of $R_{\max}$ on the simulated hysteresis loop is not pronounced.

It should be noted that due to the special construction (or load distribution characteristics) of bolted connection, the contact radius of smooth surfaces may be

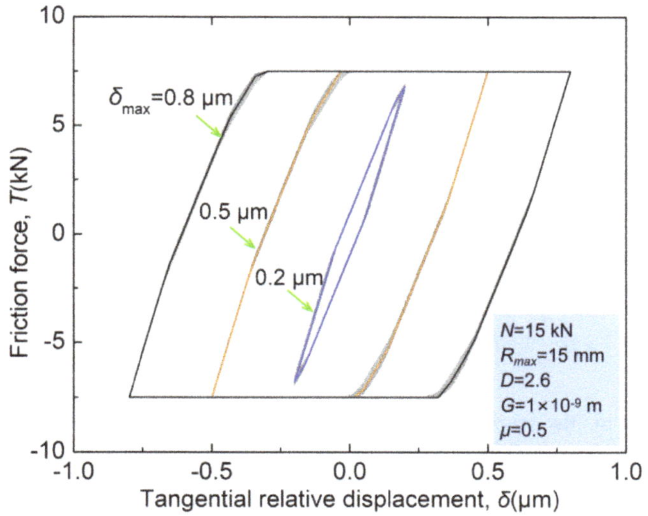


**Figure 7.15.** Force–displacement curves under different displacement amplitudes and contact radius $R_{\max}$ (with 20% dispersion).

close to the radius of rough surfaces (but the actual contact area may be much different). Therefore, the simplification of the finite element model is feasible for contact radius estimation.

Furthermore, a finite element model considering the surface roughness layer is built to identify the contact radius using the method in [39, 41]. The rough surface profile is reconstructed based on the identified fractal parameters and the W–M function. However, due to the large scale span feature of the rough surface (from about 0.1 $\mu$m to 30 mm) and the limitation of computer performance (ThinkStation P920, Intel Xeon Platinum 8163 CPU @2.50 GHz, RAM 128 G), it is impossible to perform full-scale contact analysis using the finite element method. Therefore, only a small contact portion with dimensions of 1 mm $\times$ 1 mm is modeled for contact analysis, as shown in figure 7.16, and the contact between the two rough deformable surfaces is simplified to the contact between a rough deformable surface and a smooth rigid surface.

The entire interface is treated as a collection of many contact portions. The normal load applied to the contact bodies is decreased linearly along the radial direction of the joint interface. If the obtained contact area under a certain normal load is less than 1% of the nominal contact area (1 mm$^2$), the contact is considered a gap (no contact). The resulting contact radius (from 14 to 15 mm for $N_0$=14 kN) by the rough surface finite element model is close to that (14.7 mm) by the smooth surface finite element model. Figure 7.17 shows the finite element model considering a rough surface and the contact area under different normal loads. However, limited by computer performance, this method only can implement a rough estimation of the contact radius.

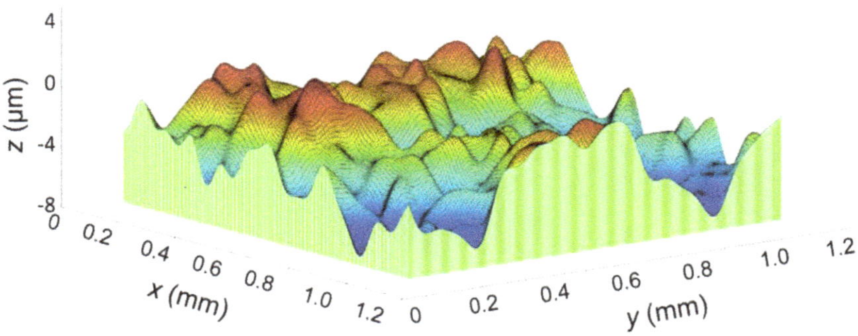


**Figure 7.16.** Reconstructed rough contact surface (dimensions of 1 mm × 1 mm).

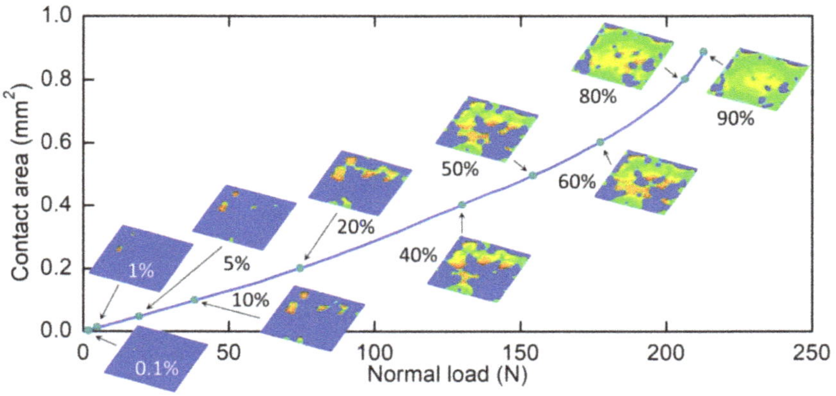


**Figure 7.17.** Simulated contact area as a function of normal load.

### 7.3.4 Effect of interface discretization

This section evaluates the impact of the number of divided sub-regions (or the Jenkins elements) on the model prediction performance over a wide range of displacement amplitudes. A simple case with only one contact surface is considered. The model parameters are as follows: contact radius $R_{\max} = 15$ mm, fractal dimension $D = 2.6$, fractal roughness $G = 1 \times 10^{-9}$, friction coefficient $\mu = 0.5$, and bolt preload $N_0 = 15$ kN.

Figures 7.18 and 7.19 show the equivalent real stiffness $k_R$ and the imaginary stiffness $k_I$ versus the relative displacement amplitude. Models with different numbers ($h = 1$, 4, 7, and 10) of the divided sub-regions are compared. As the number of $h$ increases, the differences in the equivalent stiffness predicted by different models become smaller and smaller. When the relative displacement amplitude is small, this difference is quite noticeable. At this point, the force–displacement curve predicted by

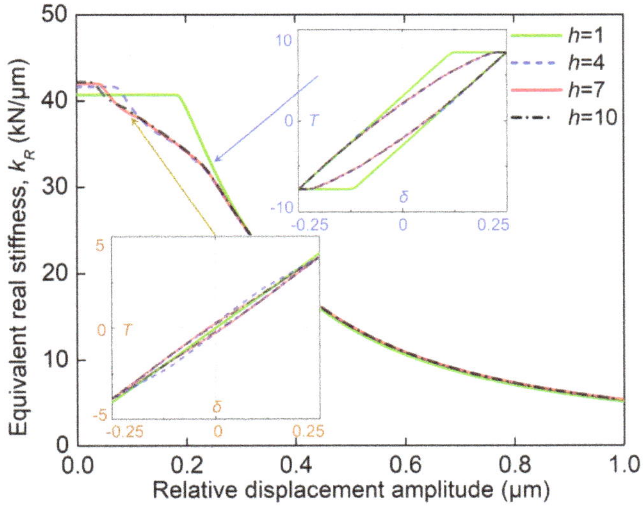


**Figure 7.18.** Equivalent real stiffness as a function of relative displacement amplitudes for the different divided sub-regions (or the Jenkins elements). (Reproduced with permission from [17]. Copyright 2022 Elsevier.)

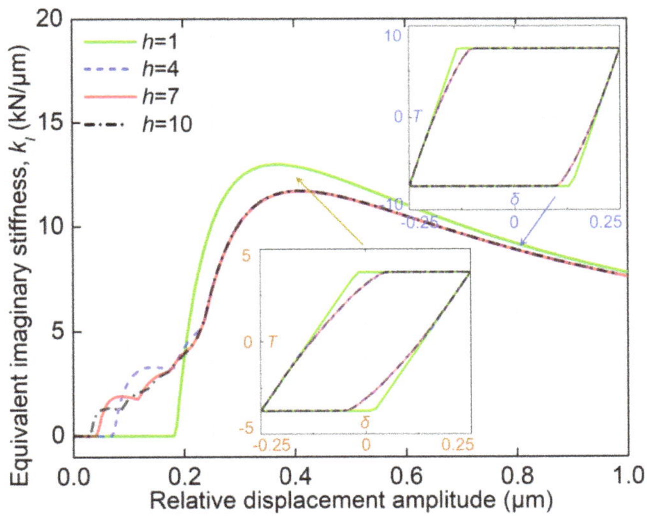


**Figure 7.19.** Equivalent imaginary stiffness as a function of relative displacement amplitudes for the different divided sub-regions (or the Jenkins elements). (Reproduced with permission from [17]. Copyright 2022 Elsevier.)

the model with a larger $h$ still forms a loop, while the force–displacement curve obtained by the model with a smaller $h$ is a straight line (indicating no hysteresis), as shown in the inset of figure 7.18. For larger displacement amplitudes (>0.1 $\mu$m), the differences in the predicted equivalent stiffness among the different models are barely visible. Considering both model complexity and prediction accuracy, it can be concluded that the model with $h = 4$ is suitable for characterizing the frictional hysteresis behavior of bolted joint structures.

## 7.4 Summary

This chapter introduces a multi-scale friction modeling method that combines a physics-based rough contact model with a discretized Iwan model. This approach overcomes the complexity which usually requires a model of the dynamic behavior of the joint and the frequency response measured with a dedicated fretting test rig. The main content and conclusions are as follows.

The above friction model comprehensively considers the multi-scale characteristics of randomly rough contact surfaces, the non-uniform distribution of contact pressure, and the convenience of phenomenological contact models in simulating frictional hysteresis. This method is implemented based on fractal contact theory and the discretized Iwan model. The fractal contact theory is used to calculate the tangential contact stiffness of the interface, and the interface is divided into several sub-regions to account for the distribution of contact pressure. A discretized Iwan model, composed of multiple Jenkins elements with different contact stiffnesses and critical slip forces, is then established to describe frictional hysteresis.

In terms of experiments, a novel test device is introduced, with the joint specimens arranged symmetrically to measure the frictional force at the bolted joint interface as purely as possible. Meanwhile, to avoid the influence of additional bending moments generated during installation on the contact conditions of the joint interface, the fixed specimen is designed with high flexibility while ensuring the overall buckling stability of the assembly. The experimental hysteresis loops indirectly confirms the theoretical speculation in [17] regarding the source of residual stiffness.

Finally, it is worth emphasizing that in fractal contact analysis, the modeling of initial tangential contact stiffness does not depend on the friction coefficient. Therefore, a physics-based friction coefficient model could potentially be integrated into this method as a separate module in the future.

Parts of this chapter have been reprinted with permission from [17].

## Bibliography


[1] Eriten M, Polycarpou A A and Bergman L A 2011 Physics-based modeling for fretting behavior of nominally flat rough surfaces *Int. J. Solids Struct.* **48** 1436–50

[2] Kang H, Li Z M, Liu T, Zhao G, Jing J and Yuan W 2021 A novel multiscale model for contact behavior analysis of rough surfaces with the statistical approach *Int. J. Mech. Sci.* **212** 106808

[3] Wang R, Zhu L and Zhu C 2017 Research on fractal model of normal contact stiffness for mechanical joint considering asperity interaction *Int. J. Mech. Sci.* **134** 357–69

[4] Shi W and Zhang Z 2022 Contact characteristic parameters modeling for the assembled structure with bolted joints *Tribol. Int.* **165** 107272

[5] Liu J, Ma C, Wang S, Wang S and Yang B 2019 Contact stiffness of spindle-tool holder based on fractal theory and multi-scale contact mechanics model *Mech. Syst. Sig. Process.* **119** 363–79

[6] Yu X, Sun Y and Wu S 2022 Multi-stage contact model between fractal rough surfaces based on multi-scale asperity deformation *Appl. Math. Modell.* **109** 229–50
[7] Armand J, Salles L, Schwingshackl C W, Süß D and Willner K 2018 On the effects of roughness on the nonlinear dynamics of a bolted joint: a multiscale analysis *Eur. J. Mech.* A **70** 44–57
[8] Abad J, Medel F J and Franco J M 2014 Determination of Valanis model parameters in a bolted lap joint: experimental and numerical analyses of frictional dissipation *Int. J. Mech. Sci.* **89** 289–98
[9] de Wit C C, Olsson H, Astrom K J and Lischinsky P 1995 A new model for control of systems with friction *IEEE Trans. Autom. Control* **40** 419–25
[10] Brake M R W 2017 A reduced Iwan model that includes pinning for bolted joint mechanics *Nonlinear Dyn.* **87** 1335–49
[11] Li D, Botto D, Xu C and Gola M M 2020 A new approach for the determination of the Iwan density function in modeling friction contact *Int. J. Mech. Sci.* **180** 105671
[12] Li D, Xu C, Kang J and Zhang Z 2020 Modeling tangential friction based on contact pressure distribution for predicting dynamic responses of bolted joint structures *Nonlinear Dyn.* **101** 255–69
[13] Li C, Jiang Y, Qiao R and Miao X 2021 Modeling and parameters identification of the connection interface of bolted joints based on an improved micro-slip model *Mech. Syst. Sig. Process.* **153** 107514
[14] Chen J, Zhang J, Hong J and Zhu L 2019 Modeling tangential contact of lap joints considering surface topography based on Iwan model *Tribol. Int.* **137** 66–75
[15] Yang H, Xu C and Guo N 2023 Modelling tangential friction considering contact pressure distribution of rough surfaces *Mech. Syst. Sig. Process.* **198** 110406
[16] Yang H, Li D, Sun J and Xu C 2024 Multiscale modeling of friction hysteresis at bolted joint interfaces *Int. J. Mech. Sci.* **282** 109586
[17] Li D, Botto D, Li R, Xu C and Zhang W 2022 Experimental and theoretical studies on friction contact of bolted joint interfaces *Int. J. Mech. Sci.* **236** 107773
[18] Ahmadian H and Jalali H 2007 Identification of bolted lap joints parameters in assembled structures *Mech. Syst. Sig. Process.* **21** 1041–50
[19] Yuan P P, Ren W X and Zhang J 2019 Dynamic tests and model updating of nonlinear beam structures with bolted joints *Mech. Syst. Sig. Process.* **126** 193–210
[20] Gimpl V, Fantetti A, Klaassen S W B, Schwingshackl C W and Rixen D J 2022 Contact stiffness of jointed interfaces: a comparison of dynamic substructuring techniques with frictional hysteresis measurements *Mech. Syst. Sig. Process.* **171** 108896
[21] Umer M and Botto D 2019 Measurement of contact parameters on under-platform dampers coupled with blade dynamics *Int. J. Mech. Sci.* **159** 450–8
[22] Yang D, Lu Z R and Wang L 2021 Parameter identification of bolted joint models by trust-region constrained sensitivity approach *Appl. Math. Modell.* **99** 204–27
[23] Jalali H, Ahmadian H and Mottershead J E 2007 Identification of nonlinear bolted lap-joint parameters by force-state mapping *Int. J. Solids Struct.* **44** 8087–105
[24] Marshall M B, Lewis R and Dwyer-Joyce R S 2006 Characterisation of contact pressure distribution in bolted joints *Strain* **42** 31–43
[25] Stephen J T, Marshall M B and Lewis R 2014 An investigation into contact pressure distribution in bolted joints *Proc. Inst. Mech. Eng.* C **228** 3405–18

[26] Yan W and Komvopoulos K 1998 Contact analysis of elastic-plastic fractal surfaces *J. Appl. Phys.* **84** 3617–24
[27] Johnson K L 1987 *Contact Mechanics* (Cambridge: Cambridge University Press)
[28] Kogut L and Etsion I 2003 A finite element based elastic–plastic model for the contact of rough surfaces *Tribol. Trans.* **46** 383–90
[29] Lin L P and Lin J F 2005 An elastoplastic microasperity contact model for metallic materials *ASME J. Tribol.* **127** 666–72
[30] Chang W R, Etsion I and Bogy D B 1987 An elastic–plastic model for the contact of rough surfaces *ASME J. Tribol.* **109** 257–63
[31] Mindlin R D and Deresiewicz H 1953 Elastic spheres in contact under varying oblique forces *ASME J. Appl. Mech.* **20** 327–44
[32] Li D, Xu C, Botto D, Zhang Z and Gola M M 2020 A fretting test apparatus for measuring friction hysteresis of bolted joints *Tribol. Int.* **151** 106431
[33] Li D, Xu C, Li R and Zhang W 2022 Contact parameters evolution of bolted joint interface under transversal random vibrations *Wear* **500** 204351
[34] Lavella M, Botto D and Gola M M 2013 Design of a high-precision, flat-on-flat fretting test apparatus with high temperature capability *Wear* **302** 1073–81
[35] Zhao B, Wu F, Sun K, Mu X, Zhang Y and Sun Q 2021 Study on tangential stiffness nonlinear softening of bolted joint in friction-sliding process *Tribol. Int.* **156** 106856
[36] Zhao B, Sun Q, Yang Y, Sun K and Liu Z 2022 Study on interface non-uniform slip of combined rotor considering real preload distribution *Tribol. Int.* **169** 107482
[37] Pan W, Li X, Wang L, Guo N and Mu J 2017 A normal contact stiffness fractal prediction model of dry-friction rough surface and experimental verification *Eur. J Mech.* A **66** 94–102
[38] Gastaldi C and Gola M M 2016 On the relevance of a microslip contact model for under-platform dampers *Int. J. Mech. Sci.* **115** 145–56
[39] Hyun S, Pei L, Molinari J F and Robbins M O 2004 Finite-element analysis of contact between elastic self-affine surfaces *Phys. Rev.* E **70** 026117
[40] Pei L, Hyun S, Molinari J F and Robbins M O 2005 Finite element modeling of elasto-plastic contact between rough surfaces *J. Mech. Phys. Solids* **53** 2385–409
[41] Thompson M K and Thompson J M 2010 Considerations for the incorporation of measured surfaces in finite element models *Scanning* **32** 183–98

**Chao Xu and Dongwu Li**

# Chapter 8

# Modeling of wear at bolted joint interfaces

This chapter presents a numerical method for the long-term dynamic analysis of bolted connection structures. The interface wear-induced mechanical behavior evolution is considered by fitting the variation of contact parameters with the cumulative dissipated energy. The fitted contact parameters functions are then integrated into the generalized Iwan model to reproduce the evolution of interface hysteresis. A single degree of freedom lumped parameter model with the interface wear is presented to show the influence of wear on the dynamic response. This work can provide the basis for the dynamic analysis of long-lasting joint structures in which wear plays a fundamental role in modifying the contact parameters.

## 8.1 Introduction

Fretting wear [1, 2] is a multi-physics and multi-scale phenomenon caused by the relative motion of contact interfaces. The authors of [3] provide an overview of fretting wear modeling methods. These research works mainly focus on fretting wear modeling in quasi-static conditions, but few studies have explored its impact on structural dynamic responses. Most wear simulations are based on finite element methods [4–7]. A typical analysis process for interface wear is shown in figure 8.1 [7], where the wear calculation is based on an energy method. After each cycle, the finite element mesh of the contact surface is updated by adjusting each node to reflect the wear depth. This process is repeated until the desired number of cycles is completed or the maximum wear depth is reached. In [8], a novel finite element modeling strategy was developed to account for the effect of the third body layer. Of course, these methods are computationally expensive.

To overcome the prohibitive computational costs of the above methods in industrial applications, Gallego *et al* [9] proposed a quasi-static multi-scale approach that couples the finite element model for global structural behavior with a semi-analytical solver. This approach allows for much finer discretization of the contact area while keeping computational cost low. Additionally, methods based on the

doi:10.1088/978-0-7503-6214-6ch8  

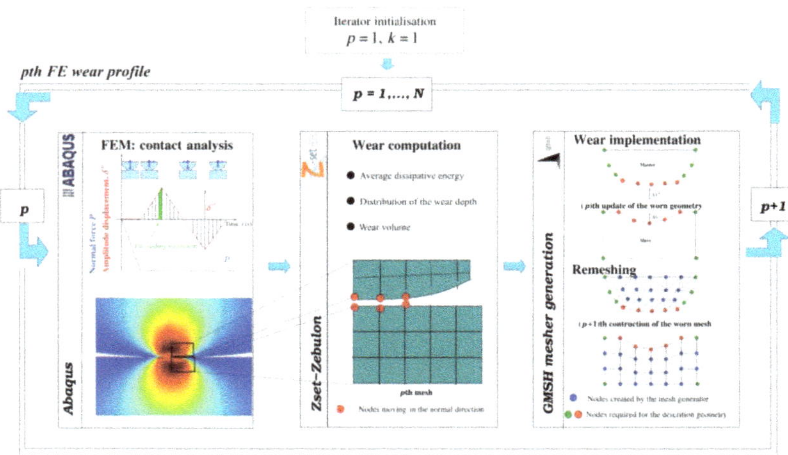


**Figure 8.1.** Schematic of wear simulation using the finite element method. (Reproduced from [7]. CC BY 4.0.)

half-space assumption [10] and numerical techniques, such as the boundary element method (BEM) proposed by Sfantos and Aliabadi [11, 12], have also been developed for wear simulation and optimization.

Since the new century, the impact of wear on the dynamic evolution of structures with frictional contact has gradually attracted the attention of scholars, especially in the field of turbine engines. Jareland *et al* [13] combined the Archard wear model with the shear layer model to study the effect of the wear behavior of the damper contact surface on the dynamic response of the mistuned blade disk. They found that as the degree of wear increases, the resonance amplitude of the blade tip first increases and then decreases. Salles *et al* [14–16] employed dual time stepping algorithms with the high-order harmonic balance method to examine wear effects on the dynamics of free-standing blades with blade root joints. They also developed a multi-scale method to address the minor changes in dynamics due to small wear amounts per fretting cycle. Armand *et al* [17, 18] extended the previous work and proposed a new multi-scale approach that operates at two different scales: the macro-scale for dynamic analysis and the micro-scale for contact analysis and wear calculation, which can better discretize the contact area and thus improve the accuracy.

Recently, Tamatam *et al* [19] implemented a coupled static/dynamic harmonic balance method with a wear energy approach and an adaptive wear logic to study the impact of wear on the nonlinear dynamic response curves and the contact interface evolution under various scenarios. Fantetti *et al* [20] performed curve fitting on the evolution of contact parameters obtained from fretting wear tests and obtained mathematical expressions for contact stiffness and friction coefficient with respect to cumulative dissipated energy. They then combined these formulas with the Bouc–Wen model to analyze the evolution of the frequency response curve of the concentrated parameter model of the test device with increasing wear. This chapter presents a similar approach to reproduce the wear-induced evolution of contact parameter and hysteresis behavior [21].

## 8.2 Wear-dependent contact parameters

Taking the wear tests of smooth contact surface in chapter 3 (tests #2 and #4) as an example, the evolution equations of contact parameters are constructed. Considering the drawbacks of polynomial fitting, such as numerous parameters and poor fitting accuracy, this paper adopts exponential basis functions for fitting and constructs a parameter evolution equation. A single curve can be fitted with at most two coefficients. The coefficients of these basis functions are obtained by fitting the experimental curves. A best fit procedure with the least square method is used.

Table 8.1 lists the constructed equations of bolt preload and wear-dependent contact parameters. The coefficients of the exponential terms in the equations are applicable to both tests #2 and #4. The parameters with a subscript of '0' represent the initial values of the fitted variables, which can be obtained from experimental data. $a_i$ $(i = 1, 2, 3)$ and $b_i$ $(i = 1, 2)$ denote the multipliers of the basis functions, and $b_i$ ranges from 0 to 1. $c$ and $d$ are the exponents of the basis function, and their selection refers to the results in [20].

Table 8.2 lists the initial values of bolt preload and contact parameters, and the coefficients of the exponential basis functions. Table 8.3 lists the multipliers of the basis functions obtained by least squares fitting, as well as their standard deviations

**Table 8.1.** Functions of bolt preload and wear-dependent contact parameters. (Reproduced with permission from [21]. Copyright 2020 Elsevier.)

| Variables | Functions |
|---|---|
| Bolt preload | $N_b(E_d) = N_{b0}[a_1 + (1 - a_1)e^{cE_d}]$ |
| Contact stiffness | $k_t(E_d) = k_{t0}\{a_2 + (1 - a_2)[b_1 e^{cE_d} + (1 - b_1)e^{dE_d}]\}$ |
| Friction coefficient | $\mu(E_d) = \mu_0\{a_3 + (1 - a_3)[b_2 e^{cE_d} + (1 - b_2)e^{dE_d}]\}$ |

**Table 8.2.** Fitted coefficients of the wear-dependent contact parameter functions. (Reproduced with permission from [21]. Copyright 2020 Elsevier.)

| Variables | Test #2 | Test #4 |
|---|---|---|
| Bolt preload | $N_{b0} = 710$ N, $c = -0.3$ | $N_{b0} = 745$ N, $c = -0.3$ |
| Contact stiffness | $k_{t0} = 105$ N $\mu$m$^{-1}$, $c = -0.3$, $d = -5$ | $k_{t0} = 115$ N $\mu$m$^{-1}$, $c = -0.3$, $d = -5$ |
| Friction coefficient | $\mu_0 = 0.3$, $c = -0.3$, $d = -5$ | $\mu_0 = 0.34$, $c = -0.3$, $d = -5$ |

**Table 8.3.** Fitted coefficients and its standard deviations with 95% confidence bounds. (Reproduced with permission from [21]. Copyright 2020 Elsevier.)

| | Values | | With 95% confidence bounds | |
|---|---|---|---|---|
| Coefficients | Test #2 | Test #4 | Test #2 | Test #4 |
| $a_1$ | 0.72 | 0.72 | ± 0.040 | ± 0.031 |
| $a_2$ | 1.71 | 1.31 | ± 0.003 | ± 0.003 |
| $a_3$ | 1.16 | 1.12 | ± 0.012 | ± 0.010 |
| $b_1$ | 0.4 | 0.8 | ± 0.015 | ± 0.020 |
| $b_2$ | 0.2 | 0.5 | ± 0.019 | ± 0.014 |

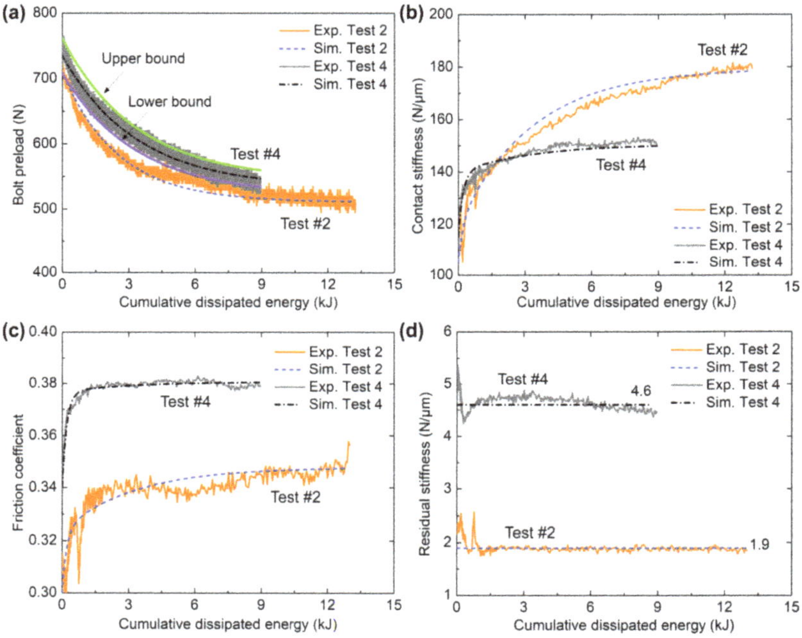


**Figure 8.2.** Measured and fitted bolt preload and contact parameters as a function of cumulative dissipated energy: (a) bolt preload, (b) tangential contact stiffness, (c) friction coefficient, and (d) residual stiffness. (Adapted with permission from [21]. Copyright 2020 Elsevier.)

within 95% confidence bounds. The results show that the fitted multipliers are consistent with the ratio of the final value to the initial value of the parameter, so they can also be directly obtained through this ratio.

In addition, the residual stiffness is considered to be independent of wear cycles because it is related to the bending stiffness of the bolt shank, which is not affected by changes in bolt preload. The results in figure 8.2(d) support this assumption.

Figure 8.2 compares the fitted contact parameters with the experimental results in tests #2 and #4. It can be seen that the two groups of results are in good agreement. Figure 8.2(a) takes the bolt preload in test #4 as an example, considering the standard deviation within 95% confidence interval of the fitting coefficient, and gives the upper and lower bounds of the fitting curve.

## 8.3 Wear model

To reproduce the evolution of interface friction hysteresis behavior with increasing wear, the authors of [21] adopt the generalized Iwan model described in chapter 6 for simulating the hysteresis behavior as the framework of the wear model, and use the above contact parameter equations as the model input, ultimately realizing the description of wear behavior. Figure 8.3 presents a flowchart of this method, which includes contact parameter equations and a description of hysteresis behavior.

Substituting the evolution equations of contact parameters into the constitutive equation of the generalized Iwan model, the wear equation can be obtained. The relationship between the friction force and displacement in the partial slip state and gross slip state is given as

$$
\begin{aligned}
T(\delta, E_d) = {} & \frac{\sqrt{\frac{\pi(6-2\pi)[k_t(E_d)+k_r(E_d)]\delta}{3\mu(E_d)N_0(E_d)} + (\pi-2)^2}}{6(\pi-3)^2} \\
& \times \left[2(\pi-3)[k_t(E_d)+k_r(E_d)]\delta - \frac{3(\pi-2)^2\mu(E_d)N_0(E_d)}{\pi}\right] \qquad (8.1) \\
& + \frac{\frac{3(\pi-2)^3\mu(E_d)N_0(E_d)}{\pi} + (3\pi^2-21\pi+36)[k_t(E_d)+k_r(E_d)]\delta}{6(\pi-3)^2},
\end{aligned}
$$

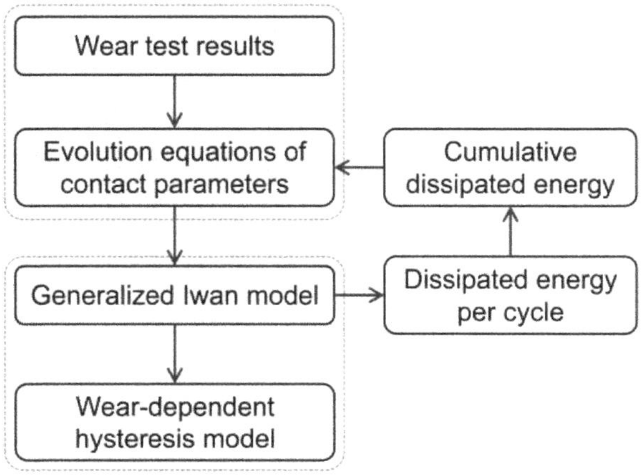


**Figure 8.3.** Flowchart for simulating the wear-dependent hysteresis.

and

$$T(\delta, E_d) = \mu(E_d)N_0(E_d) + k_r(E_d)\delta. \tag{8.2}$$

Substituting the above equations into the Masing formulas, the mathematical expression of the hysteresis curve considering the wear behavior under cyclic load can be obtained.

There are two independent variables in equations (8.1) and (8.2), i.e. the interface relative displacement $\delta$ and the cumulative dissipated energy $E_d$, with different time scales. The cumulative dissipated energy is defined within one reciprocating cycle and is a step function of time. The interface relative displacement is a continuous function of time. In the process of calculating the hysteresis curve, the step size of the cumulative dissipated energy is one cycle.

After each motion cycle, the cumulative dissipated energy needs to be recalculated. Then, the contact parameters for the next cycle are updated according to the cumulative dissipated energy. By repeating this process, the hysteresis curve that accounts for wear evolution can be obtained. Figure 8.4 shows the evolution of the hysteresis loops with increasing wear simulated by the above method.

To evaluate the effectiveness of the above wear model, a set of simulated hysteresis curves are compared with experimental results. Figure 8.5 presents a comparison between the measured and simulated hysteresis curves at four wear moments, with a dissipated energy interval of 2 kJ between each two wear moments. It can be seen that in all four cases, the wear model can accurately reproduce the experimental results.

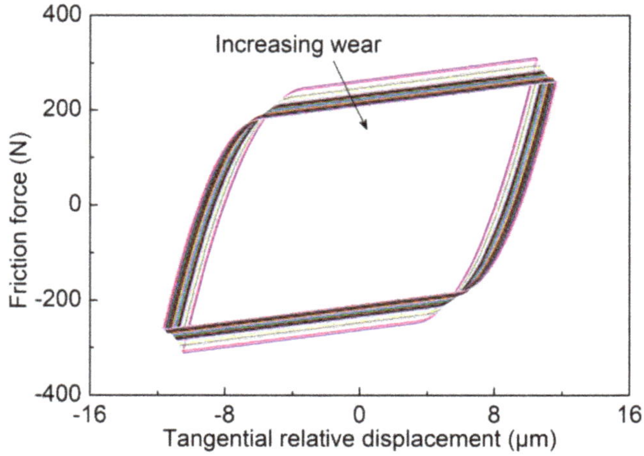


**Figure 8.4.** Evolution of the simulated hysteresis curves with increasing wear. (Reproduced with permission from [21]. Copyright 2020 Elsevier.)

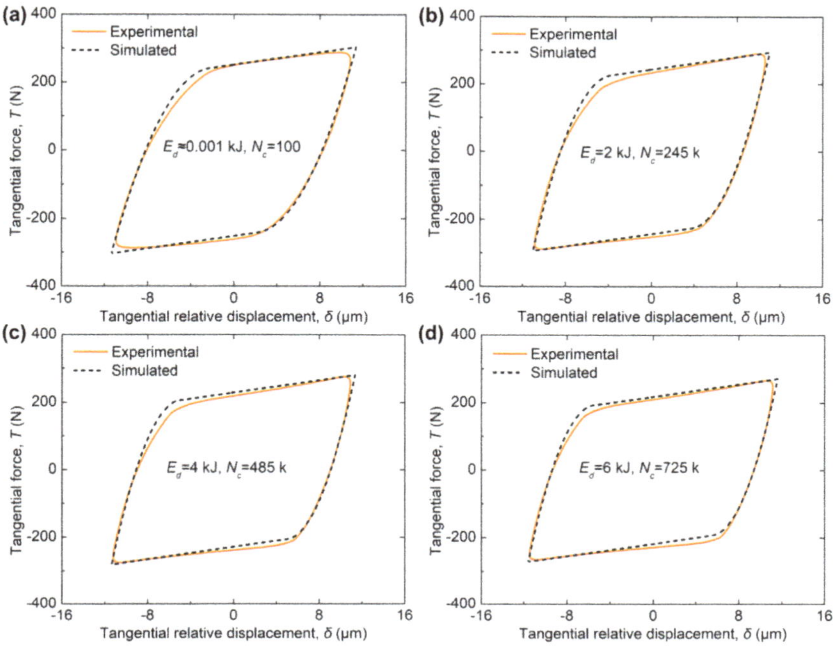


**Figure 8.5.** Comparison of the measured and simulated hysteresis curves in test #4, where $E_d$ and $N_c$ present the cumulative dissipated energy and wear cycle number, respectively: (a) $E_d = 0$, $N_c = 100$, (b) $E_d = 2$ kJ, $N_c = 245$ k, (c) $E_d = 4$ kJ, $N_c = 485$ k, and (d) $E_d = 6$ kJ, $N_c = 725$ k. (Adapted with permission from [21]. Copyright 2020 Elsevier.)

## 8.4 Influence of wear on the dynamic response

In this section, the wear model is integrated into a single degree of freedom lumped parameter model of bolted connections to simulate the impact of interface wear on the dynamic response of joint structures. The model is shown in figure 8.6. The governing equation of this system is written as

$$m\ddot{\delta}(t) + c_v\dot{\delta}(t) + k\delta\,(t) + f_{nl}\,(\delta,\,\dot{\delta},\,E_d) = F_e(t), \tag{8.3}$$

where the model parameters include: lumped mass $m = 5$ kg, damping coefficient $c_v = 0.1\ \mathrm{N \cdot s\,m^{-1}}$, and stiffness $k = 120\ \mathrm{N}\,\mu\mathrm{m}^{-1}$, and the external excitation is $f_0\cos(\omega t)$.

The harmonic balance method is used to solve the dynamic response of the system. Figure 8.7 plots the frequency response curves at different excitation amplitudes and wear stages. The results show that when the excitation amplitude is 0.1, 1 and 10 N, the influence of interface wear on the structural dynamic response is pronounced.

For the same excitation amplitude, as the interface wear aggravates (with the increase in cumulative dissipated energy), the resonance peak of the frequency response curve gradually shifts to higher frequencies, while the peak value shows a decreasing trend. Throughout the wear period, the resonance frequencies of the

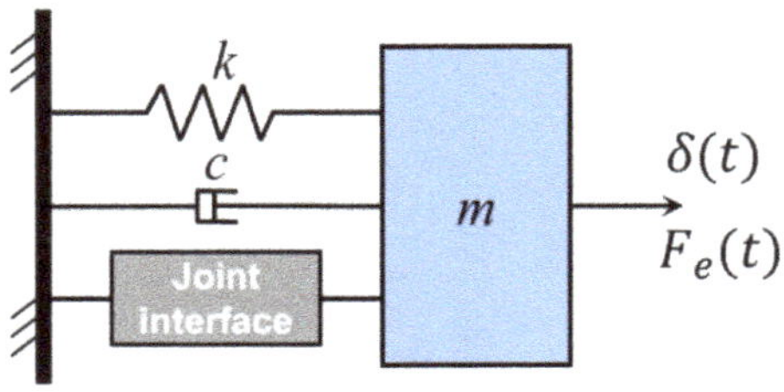


**Figure 8.6.** A single degree of freedom lumped parameter model of bolted connection structure considering interface wear.

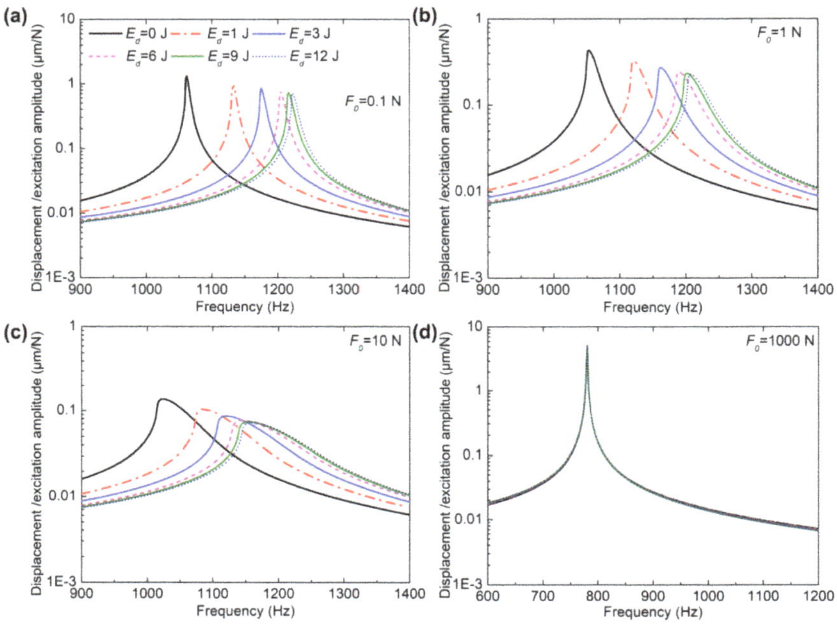


**Figure 8.7.** Frequency response curves at different interface wear stages and different excitation amplitudes: (a) $F_0 = 0.1$ N, (b) $F_0 = 1$ N, (c) $F_0 = 10$ N, and (d) $F_0 = 1000$ N.

system in the first three conditions increase by 15%, 15%, and 13%, and the resonance peaks decrease by 47%, 47%, and 50%.

Moreover, it can be seen that in the early stages of wear, the changes in the resonance frequency and peak of the frequency response curve are more drastic, while in the later stages of wear, they gradually stabilize, which is consistent with the evolution trend of contact parameters. For the case of an excitation amplitude of 1000 N, the differences in the frequency response curves at various wear stages are almost negligible, as the gross slip regime dominates the interface motion and the system can be considered an approximately linear system.

In summary, except for the case of large external excitation, the impact of interface wear behavior on the dynamic characteristics of connection structures is significant, particularly the shift in resonance frequency, which increases the difficulty of controlling the system. Therefore, in certain mechanical systems with severe interface wear, it is necessary to consider its effects on the dynamic characteristics of the system.

## 8.5 Summary

This chapter presents a method to simulate the wear behavior at bolted connection interfaces by combining the experimental results in chapter 3 and the modeling approach in chapter 5. This method fits the evolution curves of contact parameters during the wear process, establishes their mathematical expressions related to cumulative dissipated energy, and integrates them with the generalized Iwan model to reproduce the evolution of hysteresis behavior.

The simulated and measured hysteresis loops are in good agreement and prove the reliability of the proposed method. It should be noted that the proposed wear-dependent contact parameters can also be combined with other contact models. The developed method can be used to simulate the dynamics of bolted joint structures in which fretting wear process heavily alters the contact conditions.

## Bibliography


[1] Archard J F 1953 Contact and rubbing of flat surfaces *J. Appl. Phys.* **24** 981–8

[2] Waterhouse R B 1984 Fretting wear *Wear* **100** 107–18

[3] Vakis A I, Yastrebov V A, Scheibert J, Nicola L, Dini D, Minfray C and Ciavarella M 2018 Modeling and simulation in tribology across scales: an overview *Tribol. Int.* **125** 169–99

[4] Podra P and Andersson S 1999 Simulating sliding wear with finite element method *Tribol. Int.* **32** 71–81

[5] McColl I R, Ding J and Leen S B 2004 Finite element simulation and experimental validation of fretting wear *Wear* **256** 1114–27

[6] Basseville S, Niass M, Missoum-Benziane D, Leroux J and Cailletaud G 2019 Effect of fretting wear on crack initiation for cylinder-plate and punch-plane tests *Wear* **420** 133–48

[7] Basseville S, Missoum-Benziane D and Cailletaud G 2023 3D finite element analysis of a two-surface wear model in fretting tests *Friction* **11** 2278–96

[8] Arnaud P, Fouvry S and Garcin S 2017 A numerical simulation of fretting wear profile taking account of the evolution of third body layer *Wear* **376** 1475–88

[9] Gallego L, Fulleringer B, Deyber S and Nelias D 2010 Multiscale computation of fretting wear at the blade/disk interface *Tribol. Int.* **43** 708–18

[10] Gallego L, Nelias D and Jacq C 2006 A comprehensive method to predict wear and to define the optimum geometry of fretting surfaces *J. Tribol.-Trans. ASME* **128** 476–85

[11] Sfantos G K and Aliabadi M H 2006 Wear simulation using an incremental sliding boundary element method *Wear* **260** 1119–28

[12] Sfantos G K and Aliabadi M H 2006 Application of BEM and optimization technique to wear problems *Int. J. Solids Struct.* **43** 3626–42

[13] Jareland M H and Csaba G 2000 Friction damper mistuning of a bladed disk and optimization with respect to wear *Turbo Expo: Power for Land, Sea, and Air* (New York: American Society of Mechanical Engineers) p 78576 V004T03A009

[14] Salles L, Blanc L, Thouverez F, Gouskov A M and Jean P 2012 Dual time stepping algorithms with the high order harmonic balance method for contact interfaces with fretting-wear *J. Eng. Gas Turbines Power-Trans. ASME* **134** 032503

[15] Salles L, Gouskov A M, Blanc L, Thouverez F and Jean P 2010 Dynamic analysis of fretting-wear in joint interface by a multiscale harmonic balance method coupled with explicit or implicit integration schemes *Turbo Expo: Power for Land, Sea, and Air* (New York: American Society for Mechanical Engineers) pp 1003–13

[16] Salles L, Blanc L, Thouverez F, Gouskov A M and Jean P 2009 Dynamic analysis of a bladed disk with friction and fretting-wear in blade attachments *Turbo Expo: Power for Land, Sea, and Air* (New York: American Society for Mechanical Engineers) pp 465–76

[17] Armand J, Pesaresi L, Salles L and Schwingshackl C W 2017 A multiscale approach for nonlinear dynamic response predictions with fretting wear *J. Eng. Gas Turbines Power-Trans. ASME* **139** 022505

[18] Armand J 2018 *Nonlinear Dynamics of Jointed Structures: A Multiscale Approach to Predict Fretting Wear and its Effects on the Dynamic Response* (London: Imperial College)

[19] Tamatam L R, Botto D and Zucca S 2023 A coupled approach to model wear effect on shrouded bladed disk dynamics *Int. J. Mech. Sci.* **237** 107816

[20] Fantetti A, Tamatam L R, Volvert M, Lawal I, Liu L, Salles L, Brake M R W, Schwingshackl C W and Nowell D 2019 The impact of fretting wear on structural dynamics: experiment and simulation *Tribol. Int.* **138** 111–24

[21] Li D, Botto D, Xu C and Gola M 2020 Fretting wear of bolted joint interfaces *Wear* **458** 203411

www.ingramcontent.com/pod-product-compliance
Ingram Content Group UK Ltd.
Pitfield, Milton Keynes, MK11 3LW, UK
UKHW051851100926
8556IPUK00005B/573